THE ART AND SCIENCE OF LIVESTOCK EVALUATION

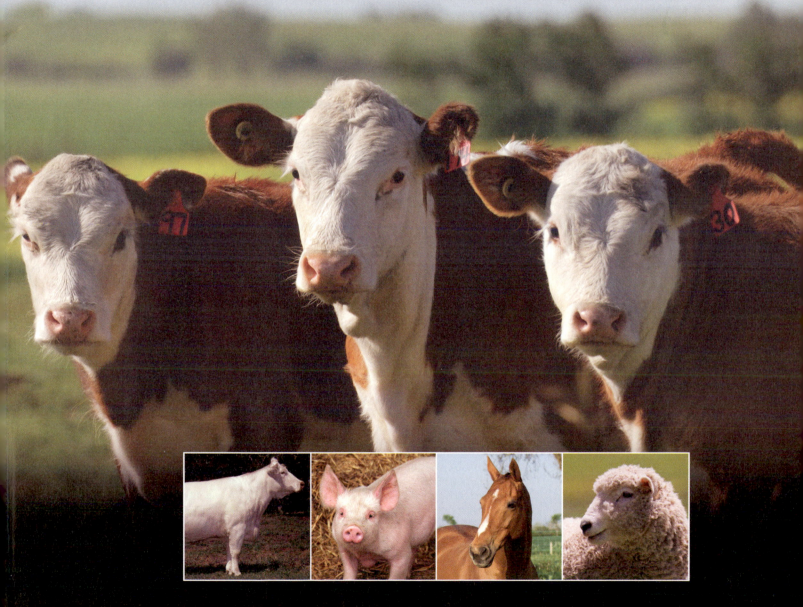

RAY V. HERREN

DELMAR
CENGAGE Learning

Australia • Brazil • Japan • Korea • Mexico • Singapore • Spain • United Kingdom • United States

The Art and Science of Livestock Evaluation
Ray V. Herren

Vice President, Career and Professional
    Editorial: Dave Garza
Director of Learning Solutions:
    Matthew Kane
Acquisitions Editor: Benjamin Penner
Managing Editor: Marah Bellegarde
Product Manager: Christina Gifford
Editorial Assistant: Scott Royael
Vice President, Career and Professional
    Marketing: Jennifer Baker
Marketing Director: Debbie Yarnell
Marketing Manager: Erin Brennan
Marketing Coordinator: Jonathan Sheehan
Production Director: Carolyn Miller
Production Manager: Andrew Crouth
Content Project Manager: Katie Wachtl
Senior Art Director: David Arsenault
Technology Project Manager: Tom Smith
Production Technology Analyst:
    Thomas Stover

For product information and technology assistance, contact us at
Cengage Learning Customer & Sales Support, 1-800-354-9706

For permission to use material from this text or product,
submit all requests online at **www.cengage.com/permissions**.
Further permissions questions can be e-mailed to
permissionrequest@cengage.com

Library of Congress Control Number: 2009927774
ISBN-13: 978-1-4283-3592-9
ISBN-10: 1-4283-3592-7

Delmar
5 Maxwell Drive
Clifton Park, NY 12065-2919
USA

Cengage Learning is a leading provider of customized learning solutions with office locations around the globe, including Singapore, the United Kingdom, Australia, Mexico, Brazil, and Japan. Locate your local office at:
international.cengage.com/region

Cengage Learning products are represented in Canada by Nelson Education, Ltd.

To learn more about Delmar, visit **www.cengage.com/delmar**

Purchase any of our products at your local college store or at our preferred online store **www.ichapters.com**

NOTICE TO THE READER
Publisher does not warrant or guarantee any of the products described herein or perform any independent analysis in connection with any of the product information contained herein. Publisher does not assume, and expressly disclaims, any obligation to obtain and include information other than that provided to it by the manufacturer. The reader is expressly warned to consider and adopt all safety precautions that might be indicated by the activities described herein and to avoid all potential hazards. By following the instructions contained herein, the reader willingly assumes all risks in connection with such instructions. The publisher makes no representations or warranties of any kind, including but not limited to, the warranties of fitness for particular purpose or merchantability, nor are any such representations implied with respect to the material set forth herein, and the publisher takes no responsibility with respect to such material. The publisher shall not be liable for any special, consequential, or exemplary damages resulting, in whole or part, from the readers' use of, or reliance upon, this material.

Printed in the United States of America
1 2 3 4 5 6 7 13 12 11 10 09

Dedication

The Art and Science of Livestock Evaluation is dedicated to Dan Rollins, Director of Feed Operations for Aviagen, North America. As cousins, we grew up together and have shared many good times and persevered through a few times that were not so good. He has always been my closest friend and best fishing buddy. His fervor for producing high quality German Short Hair bird dogs and Bonsmara cattle has always been an inspiration.

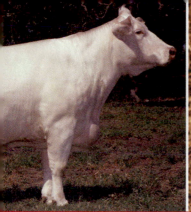

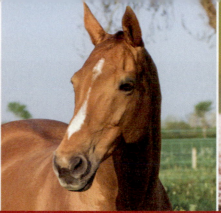

CONTENTS

SECTION THREE Cattle / 83

SECTION FOUR Sheep / 125

SECTION FIVE Horses / 181

PREFACE

Improvement of livestock through the selection process is as old as civilization. Very early on humans recognized that some animals were superior to others and began to select these animals for breeding purposes. Over the years the selection process has been perfected to both a science and an art. Livestock shows and fairs have always been a part of the agricultural scene in America. Competitive livestock judging has been around for almost as long as these agricultural fairs and livestock shows have existed. Even today as our society has moved from a rural society toward a more urban society, competitive livestock judging is still quite popular with students in 4 H, FFA, and students at the university level studying livestock evaluation.

The Art and Science of Livestock Evaluation is designed for students studying livestock evaluation as part of an Agricultural Education program. The main objective of a book is to provide instruction in swine, beef animal, sheep and horse evaluation from the viewpoint of the producer, the consumer, and livestock show standards. The evaluation principles presented in the text are for the most part, scientifically based. However, a certain degree of art plays a critical role in the selection process. To be able to visually recognize desirable traits takes a lot of practice and must be developed as an art. Strategies are outlined that will aid students in developing proper practice habits.

Information is also included that instructs students how to compete in livestock competitions. One of the main aims of the text is to help students develop the necessary skills to recognize positive and negative traits such as structural fitness, about a body fat, terms of muzzling, in animals. The development of these abilities will assist students in adapting to rapidly changing trends in livestock evaluation. Chapters are also included that deal with selection using performance data and how to evaluate using live observations and recorded data. The text contains sharp images and illustrations that help students visualize correct and incorrect traits of animals. Appendices include the proper use of livestock terminology and the rules and regulations for livestock competitions. Each chapter begins with a listing of student objectives and key terms that are used within the chapter. The chapters are concluded with a variety of questions that may be used for reviewing the chapter. In addition, practical activities are described that will enhance the student's grasp of the concepts presented in the chapter.

ACKNOWLEDGEMENTS

The author wishes to thank the following for their help in creating this text:

Dale Miller, National Hog Farmer Magazine for supplying many fine images;

Jenna Brown for her contributions in the section on scoring contests, the glossary, and the Instructor's Guide;

Christa Steincamp for contributing the chapter on using swine production data;

Kylee Johnson for contributing the chapters on horse judging;

Michael Williams for contributing the sheep chapters;

Frank Flanders for helping make several of the images;

Matt Spangler for his help in contributing materials and for editing.

SECTION ONE

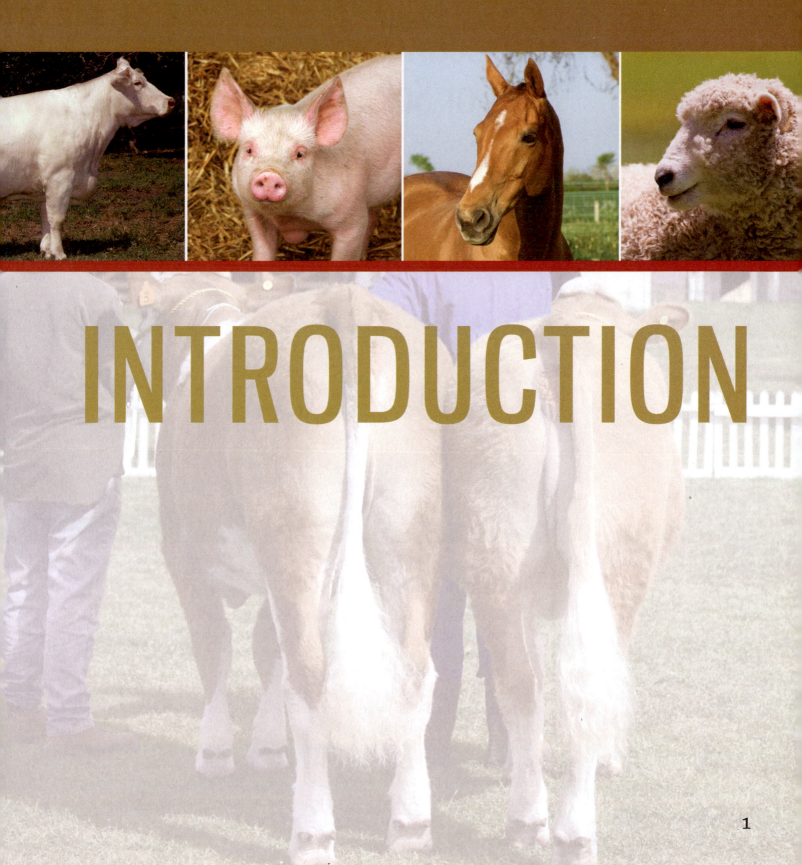

INTRODUCTION

CHAPTER 1
Selecting Livestock: Art or Science?

KEY TERMS

- Natural selection
- Breeding program
- Science
- Art
- Production data
- Binomial system
- Breeds
- Reproductive efficiency
- Growthability
- Efficiency

OBJECTIVES

As a result of studying this chapter, students should be able to:
- Describe the beginnings of livestock evaluation
- Distinguish between art and science
- Describe the different ways to classify animals
- Describe the basic traits desirable in modern livestock
- Explain how competitive livestock evaluation can help develop life skills

ORIGINS OF ANIMAL SELECTION

Anthropologists know from cave drawings that livestock have been domesticated for as long as 20,000 years, figure 1-1. Domestication came about as people began to settle in villages rather than wander from place to place searching for food. They realized that it was much easier to raise animals for food than it was to hunt them in the wild. Also, domesticated herds of animals provided a steady and readily available source of food and other products. As the raising of livestock developed, humans began to find other uses such as clothing and animals for work. Early people probably realized that the type of animals they needed were the result of reproduction of parent animals. They also realized that the offspring resembled their parents. This brought about the first efforts at animal selection. Animal producers throughout history have selected

FIGURE 1–1 Anthropologists know from cave drawings that livestock have been domesticated for as long as 20,000 years. *(Image copyright Jan Derksen, 2009. Used under license from iStockphoto.com.)*

animals based on what they thought were the traits that would improve the next generation of animals and that would better serve their needs.

In the wild state, animals developed traits that would help them survive in their environment. Animals having traits that aided them in survival stood a better chance to live and reproduce than animals without the traits. For example, cattle with the longest horns, the thickest hide, the fiercest nature, and fastest speed had a better chance at survival than those with small horns, a thin hide, and a docile nature. Wild animals needed these characteristics to defend themselves and fight off predators. Only the strongest of animals survive in the wild. Those animals that do survive breed and pass their characteristics on to the next generation. This process is known as **natural selection**. In domestication, animals no longer need many of the characteristics that increase their chance of survival in the wild. On the contrary, many of the traits that were essential for survival in the wild were a great disadvantage to animals in domestication. For instance, the fierce nature of a wild pig is far less desirable than the docile nature of the domesticated pig, figure 1-2. With humans caring for them they no longer needed long, sharp tusks to defend themselves; and in domestication, such tusks would be a danger to humans caring for them. Cattle no longer need to be able to run swiftly or to possess long horns for defense. These traits have been bred out of domesticated agricultural animals.

Over the years humans have attempted to produce, through selective breeding, those animals that do well in a domesticated state. Obviously, the conditions under which the animals are to be raised (the environment) dictate the characteristics that an animal needs to thrive. Animals selected to be raised in moderate climates with mild seasons and adequate rainfall may have a completely different set of selection criteria than animals that must survive in harsher conditions. For example, Brahman cattle are suited for hot and humid or hot and dry climates but not harsh, cold climates, figure 1-3. Cattle such as the Scottish Highlander were developed for cold climates. They would not do well in hot areas of the world.

FIGURE 1–2 With humans caring for them, animals no longer need to defend themselves with characteristics such as long, sharp tusks. *(Image copyright Andrew Hill, 2009. Used under license from iStockphoto.com.)*

FIGURE 1-3 Brahman cattle are suited for hot and humid climates, not harsh, cold climates. *(Courtesy of USDA/ARS. Photo by David Riley.)*

MODERN LIVESTOCK SELECTION

To be successful, modern livestock producers have to be able to effectively select animals that are going into either a **breeding program** or the feed lot or finishing floor. To be proficient at properly selecting livestock takes years of experience and training. This skill is truly a combination of art and science. **Science** is defined as knowledge acquired through systematic study. **Art** is the skillful use of creative imagination. Although it may seem impossible to be both, livestock evaluation can be both a science and an art. Criteria based on research such as production data can be classified as science-based selection. **Production data** are records of an animal's ancestry and may include birth weight, weaning weight, litter size, and other information. Often, desirable or undesirable characteristics such as structural soundness can be appraised only by visual evaluation. The same is true with such characteristics as muscle pattern, fat cover, and reproductive traits. To the untrained eye these can be difficult to evaluate, which is where the art factor enters into the process. It can be said that the ability to visually evaluate livestock is truly an art that must be developed.

Research has made a tremendous amount of progress in determining the type of animals to select, figure 1-4. Only within recent history has there been a truly scientific basis for the selection of animals. Most of the scientific research began around the turn of the twentieth century when experiment stations became a part of the system of land grant universities. For over 125 years, scientists have been conducting studies and gathering data on desirable traits of livestock. These traits may be those that are visual or they may be data from production records. For example, a visual trait such as the length of cannon bone in a young calf indicates the height

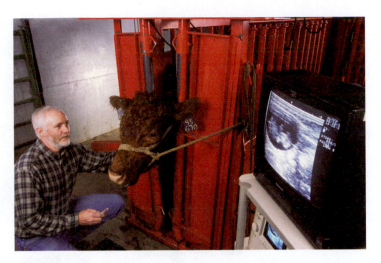

FIGURE 1–4 Scientific research has greatly advanced the selection process. *(Courtesy of USDA/ARS. Photo by Peggy Greb.)*

of the animal at maturity. This trait can be seen and proven by the collecting of data from measuring hundreds of calves' cannon bones and comparing that data to the height of the animals at maturity. Production data may involve the collecting of data such as birth weight, weaning weight, year-old weight, and carcass data.

CLASSIFYING ANIMALS

There are several ways to classify animals. As you know from studying biology, all living animals are scientifically classified using a **binomial system**. This system begins with the kingdom and ends with the genus and species of animals. For example, pigs are classified by the genus *Sus* and the species *scrofa,* so the scientific name for domestic pigs is *Sus scrofa.* Generally in livestock selection, scientific names are not of great importance to the producer or to the competitive livestock judge.

Of much more importance to producers is the purpose for which the animals are being raised. For example, sheep may be raised for either meat or wool, or both. The purpose dictates the criteria used for selecting animals. Cattle may be raised for beef or milk production. At one time there was a category called dual purpose, and these cattle were raised for both milk and beef. Today, because livestock production is highly specialized, these animals are not raised much anymore. Horses are divided into several categories. Some are raised for work such as herding cattle or pulling wagons. Some are raised for pleasure riding and some are used in competitions such as jumping, saddle events, or barrel racing, figure 1-5. Another category is racing, which is the largest spectator sport in the United States.

Animals are also classified into breeds. **Breeds** were developed because humans chose to select animals with certain characteristics for use in breeding. As breeds developed and animals bred true for the characteristics of that breed, animals were selected for desirable traits within that breed. A breed is defined as a group of animals with a common ancestry and common characteristics that breed true. *Breeding true* means that the offspring will always look like their parents. There are literally hundreds of different livestock breeds all across the world. Some are more popular whereas others are rather obscure. For hundreds of years,

FIGURE 1–5 Horses are selected for different purposes such as pleasure riding. *(Image copyright Mikhail Kondrashov, 2009. Used under license from iStockphoto.com.)*

producers have selected animals based on size, temperament, color patterns, and other physical characteristics. Only those animals meeting the proper criteria can be registered as purebreds.

Certain breeds have characteristics that are important to producers. Shorthorn cattle are known for their docile nature and even temperament. Merino sheep grow long fleeces that produce fine wool, and Suffolk sheep produce muscling that provides meat. However some breed characteristics such as color pattern have little value to the producer. For example, Hampshire hogs have a very striking pattern of black with a white stripe over the shoulder, figure 1-6. Although they look nice and are distinct, the color pattern does not provide any real benefit to the commercial producer who is much more interested in how well the animals grow and reproduce.

Livestock breeds are sometimes categorized, and each category may be evaluated in different ways. For example, cattle breeds may be designated as British breeds. These are Shorthorn, Angus, Hereford, and others that were developed in the British Isles and specifically developed to fit the conditions there. Simmental, Limousine, Charolais, and others are called continental breeds because they originated on the continental portion of Europe. A third category is referred to as Zebu cattle such as Brahman which originated in India. Similarly they were developed for a specific climate and set of conditions. Swine breeds are often categorized as sire or dam breeds. Duroc, Hampshire, and Spots are called sire breeds because of their rapid growth and muscling characteristics. Landrace and Yorkshires are called dam breeds because they have large litters and produce adequate milk for the piglets. However, most large modern swine breeding units use hybrid dams and sires that have been carefully developed from lines of several different breeds.

Horse breeds such as Tennessee Walking Horse, American Saddlebred, and Standardbreds are considered pleasure horses, whereas Thoroughbreds as race horses. Quarter Horses and Appaloosas may be considered working horses

THE ART AND SCIENCE OF LIVESTOCK EVALUATION

Join us on the web at

agriculture.delmar.cengage.com

FIGURE 1-6 Some breeds have very distinctive color markings. Hampshire pigs are black with a white stripe running from one front foot to the other. *(Image copyright Kevin Russ, 2009. Used under license from iStockphoto.com.)*

FIGURE 1-7 Draft breeds of horses are different from other breeds because of their purpose. They are raised to pull heavy loads. *(Image copyright Nicola Stratford, 2009. Used under license from iStockphoto.com.)*

used on ranches for transportation and to herd cattle. Draft breeds that are used for pulling wagons and other heavy loads include Clydesdales, Percherons, and Belgiums, figure 1-7.

Of the animals that are raised for food, there are basically two categories: those that are produced for slaughter and those that produce offspring that are raised for slaughter. There are many considerations in the selection of the type of animals for these categories. Consumers have to be pleased with the type of product that they find in the meat counter. In order to make a profit, the meat packer has to approve of the carcasses that are sent to the meat packing plant. The buyers want animals that will remain healthy until they reach the slaughter floor. The growers want animals that can gain weight quickly at an acceptable cost and that require a minimum of care. The breeders want an animal that can reproduce efficiently. All of these criteria make the selection of the modern animal a complicated process. There are three basic traits that are desirable in the modern agricultural animal: reproductive efficiency, growthability, and efficiency.

BASIC SELECTION TRAITS

Reproductive efficiency means that breeding animals must be selected that produce offspring at a regular rate. If animals are producing young at a steady rate, producers are more likely to make a profit than they would be if the animals produced fewer young, figure 1-8. This means that the males must be fertile (produce sufficient numbers of healthy sperm), must be healthy and aggressive breeders, and must live a long productive life. Females must be able to regularly come into estrus, conceive readily, produce an adequate number of healthy offspring, and produce enough milk to ensure that the young are weaned at an adequate size and weight.

FIGURE 1–8 Animals need to reproduce at a steady rate in order for producers to make a profit. *(Courtesy of USDA/ARS. Photo by Peggy Greb.)*

Growthability refers to an animal's ability to grow rapidly. The faster an animal grows, the more likely the producer is to make a profit from growing the animal. This trait is inherited from its parents and is greatly influenced by the type of care offered by the producer.

Efficiency is the ability of the animal to gain on the least amount of feed and other necessities. The producer sees that the animals are well cared for and fed. Those animals that gain the most on the least amount of feed are more desirable, figure 1-9. If one steer can gain 1 pound for every 9 pounds of feed it consumes, and another steer gains 1 pound of body weight for every 8.5 pounds of feed it consumes, then the steer that required less feed per pound of gain is said to be more efficient.

These characteristics have always been the important traits that producers have wanted. In years past, producers had a much more difficult time in predicting which animals would possess these traits. Now as a result of modern research, producers are able to predict with much more accuracy which animals will possess the desired traits. For example, research has shown that there are certain physical characteristics of animals that will predict the reproductive capability of an animal. Breed associations have developed a bank of data on the performance of offspring of a particular sire or dam. This has been brought about through the use of artificial insemination and embryo transfer. Prior to that, a dam or sire could only produce a very limited number of offspring. For example, in natural breeding, a cow usually has only one calf per year, but with embryo transfer she can have dozens of offspring each year. A bull might be able to breed only a few dozen cows in a year through natural means, but with artificial insemination he may be able to sire literally hundreds of calves per year. This has allowed the collection of a massive amount of production data for animals. As will be discussed later in this text, performance data have been compiled into values that indicate an animal's usefulness as a breeding animal.

FIGURE 1–9 Animals that can gain the most on the least amount of feed are most efficient and most desirable. *(Courtesy of USDA/ARS. Photo by Brian Prechtel.)*

COMPETITIVE LIVESTOCK EVALUATION

Competitive livestock judging is an event that has been around for quite a long time. High school programs that taught agriculture were federally funded back in 1917. About this time schools began to compete in livestock judging contests. In fact, it became so popular that schools began to look for ways to compete on the regional and then on the national level. In the 1920s, students from different states all across the country gathered in Kansas City to compete in livestock evaluation. The competition was so successful that a nationwide organization called the Future Farmers of America was organized. Today the organization is known as the National FFA Organization, and the Livestock Judging Career Development Event is still one of the most popular of all the activities of the organization, figure 1-10.

The livestock judging event has remained popular for several reasons. First, the competition teaches students to evaluate livestock based on the characteristics they see as they observe live animals. This includes developing observation skills that teach you to carefully scrutinize animals. This skill will translate into other areas as well and will help you learn to observe with a critical eye. The National FFA Organization lists the following objectives for the event:

1. To understand and to interpret the value of performance data based on industry standards

2. To measure the students' knowledge in the following categories:
    a. Making accurate observations of livestock
    b. Determining the desirable traits in animals
    c. Making logical decisions based on these observations
    d. Discussing and defending their decisions for placement
    e. Instilling an appreciation for desirable selection, management, and marketing techniques

3. To develop the ability to select and market livestock that will satisfy consumer demands and provide increased economic returns to producers; to provide positive economic returns to producers as well as meet the needs of the industry

4. To become proficient in communicating in the terminology of the industry and the consumer

FIGURE 1–10 Livestock judging competitions have remained strong in FFA since the organization began in the 1920s. *(Courtesy of Georgia Department of Education.)*

5. To identify the criteria used in grading livestock—scenarios will be used in the selection process

6. To provide an opportunity for participants to become acquainted with professionals in the industry

SUMMARY

Competitive livestock judging events may be held at the chapter, area or district, state or national levels. Each level involves several components and these may vary with the level. The top level is the National FFA Livestock Judging Career Development Event held at the national convention each year. Outlined in this chapter are the various components of the national event. A local chapter or state event may or may not include all of the components.

Livestock judging will also teach you how to critically think and analyze based on your observations. These are skills that will be used all your life in many situations. As you finish high school and go on to college or enter the workforce, you will have to make many decisions that will have a large impact on your life. Learning to problem solve and to make decisions is one of the most important skills you can develop.

STUDENT LEARNING ACTIVITIES

1. Make a list of all the ways you can think of that animals are classified. Explain the ways animals might be evaluated based on the classifications.

2. Select at least one breed of all the species of agricultural animals. List all the characteristics that distinguish that breed. Divide the characteristics as to whether or not the characteristic is of value in producing the animals.

3. Go to the internet and research the origins of a species of animal. Find out where the animal originated, when it was domesticated, and what wild characteristics had to be bred out.

FILL IN THE BLANKS

1. As the raising of livestock developed, humans began to find uses other than food such as _____ and animals for _____.

2. Over the years humans have attempted to produce through _____ _____, those animals that do well in a domesticated state.

3. Science is defined as knowledge acquired through _____ _____.

4. Art is the skillful use of _____ _____.

5. _____ _____ may involve the collecting of data such as birth weight, weaning weight, year-old weight, and carcass data.

6. A breed is defined as a group of animals with a _____ _____ and common characteristics that _____ _____.

7. _____ _____ means that breeding animals must be selected that produce offspring at a regular rate.

8. _____ refers to an animal's ability to grow rapidly.

9. _____ is the ability of the animal to gain on the least amount of feed and other necessities.

10. In the early 1900s, livestock judging competition was so successful that a nationwide organization called the _____ _____ _____ _____ was organized.

 MULTIPLE CHOICE

1. Evidence indicates that animals were domesticated as far back as:
   a) 20,000 years ago
   b) 100,000 years ago
   c) 2,000 years ago
   d) 5,000 years ago

2. In the wild, the strongest animals survive and reproduce. This is known as:
   a) wild selection
   b) natural selection
   c) maximum breeding
   d) evolution

3. Modern livestock are improved through a method known as:
   a) inbreeding
   b) mass breeding
   c) visual selection
   d) selective breeding

4. Production data is an example of:
   a) art
   b) art and science
   c) science
   d) none of the above

5. The best example of art in livestock selection is:
   a) visual appraisal
   b) data selection
   c) both a and b
   d) art is not used in livestock selection

6. An example of production data is:
   a) birth weight
   b) weaning weight
   c) neither a nor b
   d) both a and b

7. For the producer, scientific names are generally:
   a) most useful
   b) not very useful
   c) useful in most situations
   d) used only by the best producers

8. Two categories of animals used in producing meat are:
   a) large and small animals
   b) those raised for slaughter and those raised for breeding animals
   c) fat animals and nonfat animals
   d) meat type and nonmeat type

9. Efficiency refers to an animal's ability to:
   a) grow rapidly
   b) produce a lot of muscle for meat
   c) reproduce easily
   d) gain weight using the least amount of feed and other necessities

10. Banks of data are collected by:
    a) breed associations
    b) producers
    c) consumers
    d) the USDA

DISCUSSION

1. Discuss some of the characteristics needed in domesticated animals as opposed to wild animals.
2. List some of the considerations used in selective breeding.
3. Distinguish between the science and the art of livestock evaluation.
4. What are some of the types of data included in production data?
5. Describe some of the categories of animal breeds.
6. What are the basic traits desirable in the evaluation of most agricultural animals?
7. What effect did artificial insemination and embryo transfer have on production records?
8. What are some of the skills you learn in livestock judging that can be transferred to other areas of your life?
9. Why has livestock judging remained so popular?
10. At what levels are livestock judging contests held?

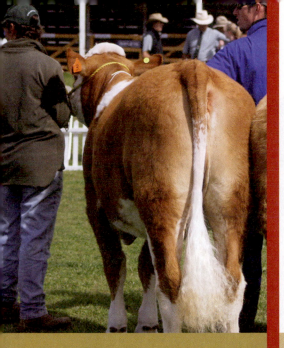

CHAPTER 2
Placing Live Classes of Animals

KEY TERMS

- Keep cull class
- Hormel system
- Official placing
- Cuts
- Hormel judging slide
- Production records

OBJECTIVES

As a result of studying this chapter, students should be able to:

- Explain the scoring system for a livestock judging contest
- Use the Hormel system of scoring
- Pon records in a judging competition

STEPS OF COMPETITIVE LIVESTOCK JUDGING

One of the first steps in competitive livestock judging is to learn how a competitive event works. First, you have to be able to critically analyze a class of four animals and rank them from top to bottom, figure 2-1. Next, you are required to analyze a set of production records that can be used along with your visual observations to correctly rank the animals according to their usefulness. You will be given a **keep cull class** where you determine which of a group of animals you wish to keep and which you think should be culled from the herd. Also, you will be given a group of market animals to grade. The most important part is that of orally communicating your rationale for placing the animals the way you did. All of these areas are used to determine your final score as a judge.

13

FIGURE 2–1 In competitive judging you will be required to rank sets of animals with the best in first place and the worst in last place. *(Courtesy of National FFA Organization.)*

HORMEL COMPUTING SYSTEM

The **Hormel system** is the most widely accepted system of scoring in the nation, and serves as the basis for calculating livestock judging scores. The Hormel system bases livestock classes on a scale of 50 points possible, and allows no option for a score of zero from a contestant. When judging, there are twenty-four different possibilities of placing a class of four animals, each noted on the livestock judging card. Most judging contests rely on the Hormel system or computer programs to help accurately and efficiently calculate contest scores. Ideally, for the Hormel system to achieve satisfactory results there should be at least two or more groups of animals for individuals to judge. Judging more than two groups allows the system to take into account the possibility of an entrant with a fluke score on one class of animals. Because the Hormel system is a consistent and fair method to calculate livestock judging classes, it is widely recognized in the United States and around the world.

THE BASIS OF THE SYSTEM

At a livestock judging contest either an individual or a group of individuals will serve as the officials for the contest, with their job being to evaluate each class and determine how each class should be placed, figure 2-2. Once the judge or judges have determined a placing for each class, this becomes the **official placing**. Each class will be broken down into three pairs which are called the top, middle, and bottom pairs. After reaching an official placing, each judge will determine how easy or difficult the decision was in placing the first animal over the second (top

FIGURE 2–2 An official judge will make the official placings. *(Courtesy of National FFA Organization.)*

pair), the second animal over the third (middle pair), and the third animal over the fourth (bottom pair). Once this decision has been made, the judge will decide on **cuts**, or the penalty for switching the top, middle, or bottom pairs in the class. For example, a cut or penalty of 1 point would indicate that the pair of animals was very similar and could easily be switched in the placing. On the other hand, a penalty of 7 points would indicate that there were no identifiable similarities between the two animals, and the pair should be obvious to the contestant.

These cuts will form the foundation for scoring the placing, and are based on the perceived degree of difficulty in choosing between pairs. While official judges are determining the values of their cuts, they must keep in mind that the total of the three cuts cannot be over 15 points for a class of four animals. If the cuts total 15 points, the middle cut cannot be larger than 5 points, and if the cuts total 14 points, the middle cut cannot be more than 8 points. If the official judges do not exceed any of these limits, it will not be possible to arrive at a final score less than zero.

COMPUTING YOUR SCORE

There are six comparisons involved in scoring a class of four animals; the number one animal should be compared to the number two animal, the number one animal should be compared to the number three animal, then one to four, figure 2-3. Next, the number two animal should be compared to the number three animal, and then the number two animal to the number four animal. Finally, a contestant should compare the number three animal to the number four animal, figure 2-4. Although this seems like a lot of decisions, there are really only three decisions to be made. This will be explained later in this chapter.

FIGURE 2–3 There are six comparisons to make in judging a class. *(Courtesy of National FFA Organization.)*

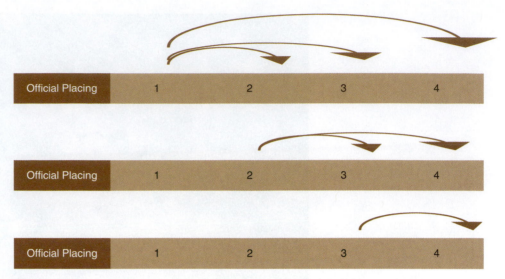

FIGURE 2–4 There are really only three decisions to make in judging a live class.

In order to fully understand the system and to be a more effective judge, you should understand how to score your card. To calculate your score it helps to create a chart comparing your placing to that of the official placing. The charts and examples below will help you in learning to compute your score manually. If your placing is the same as the official placing, you will receive the maximum 50 points possible for that class. If your placing of a pair disagrees with the official placing, points will be deducted from your score based on the cuts that have been assigned for that pair. To calculate your placing score there are three pieces of information needed: the total points possible, the official placing for the class, and the cuts for the class. You already know the maximum score possible for a class is 50 points; therefore, you need to know the official placing and the cuts for each pair in the class. For example, a contestant places a class of animals 1-3-2-4. Once the contest is complete, the contestant is informed that the official placing for the class was 1-2-3-4 with cuts of the top, middle, and bottom pairs at 2, 4, and 6 points, respectively. The contestant has only switched the middle pair in the class. Because the cut for switching the middle pair is 4 points, the contestant would deduct 4 points from 50, resulting in a score of 46, figure 2-5.

If another contestant scored this class of animals 2-1-4-3 with the official placing and cuts remaining the same, this contestant would have switched the top and the bottom pair. Remember that the cut for the top pair was 2 points, while the cut for the bottom pair was 6 points. This contestant would deduct a total of 8 points (2 + 6 = 8) from the possible 50 points, resulting in a score of 42 points for this class, figure 2-6.

The steps above work fine if only a pair is switched, but what if the placing has more than just a pair switch? Suppose another contestant has placed the same class of animals 4-3-1-2. When compared with the official placing of 1-2-3-4, note that this contestant has placed the fourth animal in first place, the third animal in second place, the first animal in third place, and the second animal in fourth place.

Official Placing	1	2	3	4
Student Placing	1	3	2	4

FIGURE 2–5 In this class the contestant switched the middle pair. Because the middle cut was 4 points, the 4 points are subtracted from 50 to give a score of 46.

Official Placing	1	2	3	4
Student Placing	2	1	4	3

FIGURE 2–6 In this class we find that the contestant has switched the top and bottom pairs according to the official placing. Because the top and bottom cuts were 2 and 6 points, respectively, we would subtract 8 (2 + 6) from 50 and arrive at a score of 42 points.

Step	Comparisons	Animal Numbers	Same as Official?	Number of Pairs Affected	Penalty
1	1st to 2nd	4 to 3	No	1	−6
2	1st to 3rd	4 to 1	No	3	−12 (2 + 4 + 6)
3	1st to 4th	4 to 2	No	2	−10(4 + 6)
4	2nd to 3rd	3 to 1	No	2	−6(2 + 4)
5	2nd to 4th	3 to 2	No	1	−4
6	3rd to 4th	1 to 2	Yes	0	0
					Total: 38

FIGURE 2–7 First decision: The contestant incorrectly placed 4 over 3 which disagrees with the official placing. A penalty of 6 points is deducted from this total score.

Second decision: The contestant incorrectly placed 4 over 1 which disagrees with the official placing. Because there is a break in all three pairs, the total loss of points is added together and deducted from the score.

Third decision: The contestant incorrectly placed 4 over 2 which disagrees with the official placing. This time only the last two pairs are broken. A penalty of 10 points is deducted from the score.

Fourth decision: The contestant incorrectly placed 3 over 1 which disagrees with the official placing. Two pairs have been broken, which results in a loss of 6 points.

Fifth decision: The contestant incorrectly placed 3 over 2 which disagrees with the official placing. One pair has been broken, which results in a loss of 4 points.

Sixth decision: The contestant has correctly placed 1 over 2. Even though it was placed in the wrong spot, no points will be deducted from the score because it matches the official placing of 1 over 2.

Because this placing will be more difficult to score, a graph can be useful to help calculate the score, figure 2-7. The total points deducted from this class will be 38, which results in a score of 12 for the contestant.

HORMEL JUDGING SLIDE

Officials or contest volunteers need an instrument to score numerous judging cards in a quick and efficient manner to maintain the time schedule of the judging contest. Because they are not afforded the time to create charts, or calculate each card manually, the **Hormel judging slide** is an essential instrument to use

FIGURE 2–8 The Hormel computing slide is often used to calculate a contestant's score.

at a contest. After the competition and the judging cards have been submitted, the Hormel computing slide may be used to score the cards, figure 2-8. The slide will allot scores from 1 to 50 according to the degree of error of each participant's indicated score.

To use, find the cuts given for the class by the judge on the bottom of the white cards. These cuts will be listed in the order of top, middle, and bottom pairs. Insert and line up the white card you have chosen in the slide so that the cuts chosen are shown in the bottom of the plastic window. Next, you will need to find the official placing order on the top of the clear plastic cards. The placing order used is listed at the top of the card. Insert the plastic card on top of the white cards in the slide and adjust so that the placing order is at the top left of the window and the scores listed on the white card appear to the right of the placing orders. The correct score will appear to the right of each possible placing order. This way, each possible placing will show a proper score next to it, which allows the officials to score quickly.

The Hormel judging slide was created and copyrighted by George A. Hormel & Company in 1975 in the United States and is a widely used instrument in judging contests. Hormel computing slides can be ordered from your county extension agent or from the National FFA Organization website. Other computer programs or websites can be used by individuals scoring judging cards. By entering the official placing, cuts for the class, and the participant's placing, individuals scoring cards for the contest can instantly be given a score for each participant's card. The Hormel system and judging slide create an efficient and effective means to score numerous judging cards, and provide a reliable and fair method to evaluate livestock in the United States. This seems like a complicated system, but it is made easier by using a scoring system called a Hormel card. This card makes scoring a lot easier and faster than hand calculating. Of course, in the larger contests such as state and national, the scoring is done by computer.

PLACING LIVE ANIMALS

During this phase of the competition, you will be required to rank four animals according to their usefulness. In order to correctly place the animals you must know how the animals are intended to be used. For example, the animals may be intended as market animals that will be slaughtered and packaged as meat. These animals would probably be placed differently than they would if they were to be used in a breeding program. There is disagreement among professional judges as to how much difference there should be between a market class and a breeding class. One side argues that market animals are intended to be slaughtered and such criteria as structural correctness does not matter. The other side counters that in a contest or show the top animals represent those that producers should be trying to produce. They contend that any animal, whether it is used as a breeding animal or is a market animal, should be structurally correct, figure 2-9. Before you begin judging the class you will be told what the animals are to be used for. For example, you will be told that the class is one of breeding gilts, a class of market hogs, a class of one-year-old breeding heifers, and so on.

Once you are given access to the class you can begin to make your placing. You should be given a front view, a side view, and a rear view of the animals. Handlers may show the animals or they may be shown in a pen loose. Either way, you should view each animal from all three perspectives. Be sure to take time when making your decision. You will have plenty of time to make up your mind on the placing. At the same time, a common mistake is taking too long and talking yourself out of your placing. Remember that the first impression is usually the best. A common mistake is for contestants to think that there is an animal or two in the class put there to confuse everyone. This is rarely the case. Most times, the animals have been carefully selected and you should be able to logically place them. Do not be fooled into thinking the class is too easy. Remember that you have practiced long and hard to be able to understand the ideal animal and the placing should be easier than you think.

FIGURE 2–9 Many official judges contend an animal should be structurally correct whether it is a breeding or a market animal. *(Courtesy of USDA/ARS. Photo by Keith Weller.)*

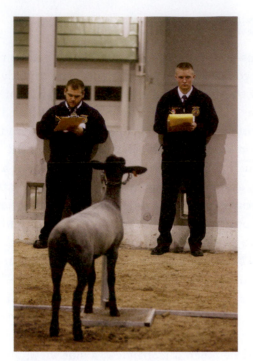

FIGURE 2–10 Develop a systematic way of analyzing the animals, one at a time. *(Courtesy of National FFA Organization.)*

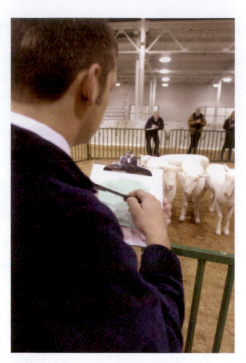

FIGURE 2–11 First select the best animal in the class and the worst animal in the class. *(Courtesy of National FFA Organization.)*

Develop a systematic way of analyzing the animals, one at a time, figure 2-10. Be sure that you look at how much muscling the animal has, how much fat, the size of bone, and relative size for the age of the animal. Also critically examine the structure of the animal and watch as the animal walks. All of these factors have to be taken into consideration before deciding the rankings of the animal. That is why the event is called judging. You have to consider all the factors and render a judgment based on your observations.

At first there seems to be a lot of decisions to make as you look at the class. However, if you approach the problem systematically you will find it to be much easier. Begin by making one decision—what is the best animal in the class? You also may start by identifying the worst animal in the class. Once you have determined the best animal, then decide which is the worst animal in the class, figure 2-11. By doing this you have selected two of the four animals in the class—the top and bottom. Then all you have to do is decide the middle pair. Do this by determining which most closely resembles your top animal and which most closely resembles your worst animal. Using this system you have arrived at your placing. Be careful about changing your mind. Change it only if you observe characteristics that you totally missed as you made your placing. Complete the class by making notes on why you placed the animals as you did. Be sure to note the name of the class such as class 1, two-year-old Angus bulls, and so forth. Develop a good system for taking notes that will help you remember the points you considered in making your placement. In the next chapter you will be given an example of how to take proper notes.

ANALYZING PRODUCTION RECORDS

On certain breeding classes you will be given a set of **production records**. For the production component, you will be given a scenario along with the production records for the class. From these data, you will be required to place the animals according to their performance, figure 2-12. A more thorough explanation of production data will be given in the chapters on judging different species. Then, you will be required to give a final placing based on a combination of data and visual observations. The following is an example of the scenario and data you will be given.

You are a commercial cattleman with a herd of mature commercial Charolais cross cows. Cattle prices are predicted to be higher for the next three years. Your management plan is to keep all of the calves and feed them out in your own feedlot. Place this class of Angus bulls based on the above criteria, visual appraisal, and performance data listed in table 2-1.

The key points in this scenario are: (1) All the calves will be put on the feedlot and not saved as replacements, therefore milk expected progeny difference (EPD) is not a consideration; and (2) growth traits are most important (weaning and yearling EPD = s) and birth weight is not a high priority because the cows are mature Charolais crosses. As will be explained in more detail in later chapters, EPD indicates how much difference in a trait, such as milk production, the offspring of an animal may be from its parents. Official placing based on performance data only is 3-2-4-1.

You would use the production records along with the live observation to place the class. The placing can be changed if you note physical attributes of the animals that make them less desirable. For example, you observe that the top place animal stands too straight on his rear legs and may have problems walking as he gets older. In this case, you may want to move him to second place and move the second place bull to the top. As mentioned, do not see more than what is there. Make sure you are correct in your observations before switching a pair.

The last thing to remember is to mark your card. This may seem like a trivial matter to mention, but many contests have been lost because a contestant failed to mark the card for a particular class. When this happens, the contestant automatically loses 50 points as the score for the class will be zero.

FIGURE 2–12 On some classes you will be given production data that are to be used in making your placings. *(Courtesy of National FFA Organization.)*

TABLE 2-1 EPD's for Angus Bulls

BULL NO.	BIRTH WEIGHT EPD	WEANING WT. EPD	MILK EPD	YEARLING WT. EPD
1	+ 1.0	+ 13.0	+ 12.0	+ 24.0
2	+ 3.0	+ 26.0	+ 7.0	+ 36.0
3	+ 4.0	+ 33.0	+ 8.0	+ 45.0
4	+ 2.0	+ 22.0	+ 5.0	+ 34.0
Angus Breed Avg.	3.0	20.0	8.0	30.0

SUMMARY

Competitive livestock judging events are scored based on a 50-point system. Points are taken off if a contestant's placing is different from that of the official judge. More points will be taken off for a mistake when judging an easy class than for a more difficult class of animals. This is known as the Hormel system. Other areas of judging will include the use of production records and oral reasons for explaining how you placed the classes. Presenting oral reasons will be covered in detail in the next chapter.

 STUDENT LEARNING ACTIVITIES

1. Make up several official placings along with cuts for the pairs and several contestant placings to go with each official placing. Use the Hormel system to calculate contestant scores.
2. Visit a livestock farm and practice placing four animals. Use the three-decision method where you decide on the top animal, the bottom animal, and which of the two middle animals is superior.
3. Practice taking notes on the placings.

 FILL IN THE BLANKS

1. One of the first steps in competitive livestock judging is to learn how a _____ _____ works.
2. The _____ _____ is the most widely accepted system of scoring in the nation, and serves as the basis for calculating livestock judging scores.
3. Once the judge or judges have determined a placing for each class, this becomes the _____ _____.
4. The judge will decide on _____, or the _____ for switching the top, middle, or bottom pairs in the class.
5. There are _____ comparisons and _____ decisions in judging a class of four animals.
6. If your placing is the same as the official judge, you will receive the maximum of _____ _____ for that class.
7. Hormel computing slides can be ordered from your _____ _____ _____ or from the _____ _____ _____ website.
8. You should be given a _____ view, a _____ view, and a _____ view of the animals in each class.
9. A common mistake is for contestants to think that there is an animal or two in the class put there to _____ _____.
10. Watch as each animal walks and critically examine the _____ of the animal.

MULTIPLE CHOICE

1. A class where you determine which animals to keep is called:
   a) contestants' choice
   b) keep cull
   c) production class
   d) reasons class

2. How many possible placings are there for a class of four animals?
   a) eight
   b) twelve
   c) twenty-four
   d) thirty-six

3. A cut or penalty of 1 point between a pair of animals indicates:
   a) it is a difficult choice
   b) it is an easy choice
   c) the animals are equal
   d) you are automatically deducted 1 point

4. The Hormel judging slide is widely used because:
   a) it is more efficient than a computer
   b) it is easy and fast to calculate a score
   c) it is the most accurate way to score a class
   d) it provides a close estimate of the score

5. When deciding how to place four animals in a class, there are:
   a) six decisions
   b) four decisions
   c) two decisions
   d) three decisions

6. When making your final score remember that:
   a) sometimes there are trick animals in the class
   b) placings are always difficult
   c) first impressions are usually the best
   d) you should change your placings at least twice

7. Final placings:
   a) should be changed at least twice
   b) should never be changed
   c) cannot be changed
   d) should be changed only if you are sure you missed a problem with an animal

8. For the production component of the contest, you will be given:
   a) only data
   b) only a scenario
   c) a scenario and production data
   d) nothing but the class of animals to observe

9. EPD stands for:
   a) expected progeny difference
   b) exact production difference
   c) extra production difficulty
   d) enough production data

10. A simple but costly mistake is:
    a) failing to place the class correctly
    b) taking too much time
    c) using the wrong score card
    d) failing to mark the score card

DISCUSSION

1. Explain what a cut is and how it is used in scoring a class of animals.
2. What is meant by the official placing?
3. Explain how to use a Hormel judging slide.
4. Discuss how you use the three-decision method to place a class of livestock.
5. What is the controversy over placing market and breeding animals the same (what are the opposing arguments)?
6. Discuss why you should be reluctant to change your initial placing of a class.

CHAPTER 3
Presenting Oral Reasons

OBJECTIVES

As a result of studying this chapter, students should be able to:

- Explain why oral reasons are such an important part of competitive judging
- Discuss how the process of presenting oral reasons works
- Take proper notes on a class of animals
- Use notes to organize a set of reasons
- Give reasons in the proper format
- Use proper terms to communicate
- List and discuss the ways reasons classes are evaluated
- Properly dress for and present an effective set of reasons

ORAL REASONS

The most important part of livestock competition is the **oral reasons** class where the contestant gets to justify how the classes are placed. Rarely will a contestant place all the class like the official judge and a few points will be deducted for switching close pairs of animals, figure 3-1. Most people in the contest will know how to place the classes and the scores on the live animal classes should be close. The oral reasons class provides a chance to make up any points lost in the live placings.

FIGURE 3–1 Rarely will a contestant place all the class like the official judge and a few points will be deducted for switching close pairs of animals. *(Courtesy of National FFA Organization.)*

Before you begin to learn about giving reasons in competitive livestock evaluation, there are two objectives you absolutely must master. The first is to be able to correctly identify the parts of an animal's body. Unless you can identify these parts, you will never be able to properly evaluate the different components that are important when evaluating an animal. Second, you must learn to talk like a livestock person. Every science has its own language with terms that are unique to that discipline. Livestock judging is no different, and to be able to effectively communicate you need to learn the necessary terms. Remember that to a degree you are judging against a standard and to properly convey your understanding of what the standard is, you must be able to talk the "language".

After these terms are learned, you need to be able to use them correctly. For example, describe the hind quarter of a breeding heifer in terms of pelvic capacity and smoothness of muscle and describe the hind quarter of a market heifer in terms of the thickness and degree of muscling. If you describe a market animal as having more pelvic capacity, you will convey that you really don't have an understanding of the animal industry. Refer to Appendix 1 to learn the parts of the animals and terminology used in livestock judging.

PRESENTING ORAL REASONS

As mentioned, perhaps the most important part of competitive judging is the ability to organize your thoughts and to present your rationale for why you placed the class as you did, figure 3-2. Remember that you are judging. This means that you are rendering an opinion based on your judgment and oral reasons give you an opportunity to justify your placings. Even professional livestock judges do not always agree on how animals should be ranked. For example, say you have a very nice pair of two-year-old Polled Hereford heifers in a class. One animal is larger

FIGURE 3-2 One of the most important parts of competitive judging is the ability to organize your thoughts and to present your rationale for why you placed the class as you did. *(Courtesy of National FFA Organization.)*

and obviously the fastest growing of the two animals. However, the other animal is almost as large but has a more feminine appearance and may be the most fertile of the pair. A judge from the southeastern United States may place the larger animal at the top of the class, whereas a judge from the western part of the United States may place the more feminine heifer at the top. Which is correct? In a sense both are correct. The southeastern judge would base the decision on personal background in a place where cattle can be managed more closely. The weather there is mild and provides adequate rainfall, so cattle are pastured in areas that can handle a cow calf unit on an acre of well-managed grass land so growth would be more important than a slight advantage in fertility, figure 3-3. In this type of management system, cattle can be watched more closely and fertility problems can be monitored.

In the West, cattle are often raised on arid, open range that might take 50 acres to support a cow calf unit. With these large areas, an advantage in growth would be outweighed by an advantage in fertility, figure 3-4. Larger cattle require more forage and would have to walk farther to obtain enough food. Also, with cattle ranging far, it is much more difficult to monitor fertility and breeding. These differences are why presenting reasons is so important.

Typically certain classes in the event will be designated as **reasons classes**. These are the classes where you should take careful notes so you can do a thorough job with presenting your reasons. You will be given two minutes to explain to the judges your rationale for your placing. The judges will want to hear reasons that indicate that you based your placing on correct observations. The judge will have seen the class and will listen for the characteristics observed in the class. Be sure to take a small notebook and a pencil to the event so you can make notes on the reasons class, figure 3-5. Keep in mind that one of the most important factors in presenting effective oral reasons is organization. If your reasons are not organized you will not be effective. The notes will help you organize your thoughts for the presentation. Also, taking notes will help you remember the important factors in

FIGURE 3–3 In the Southeast, a well-maintained pasture can support a cow and a calf on an acre of grass. *(Courtesy of NRCS. Photo by Lynn Betts.)*

FIGURE 3–4 In arid rangeland in the West, many acres of grass are required to support a cow and calf. *(Courtesy of NRCS. Photo by Tim McCabe.)*

placing the class as you did. There are several ways to organize your notes. A good system is to arrange the notes like those in table 3-1.

You will have only two minutes to present your reasons so make sure you have a systematic, concise method of presentation. This is a major reason why the first thing you have to learn is the terminology used to describe an animal. The second type of terminology you need to learn is the **comparative terms**. When

TABLE 3-1 Sample Note Taking One-year-old angus heifers Official placing: 4-2-3-1

	REASONS FOR PLACING	ADMIT/GRANT	FAULTS
Top pair	4-2 Thicker breed character Wide top Long rump	2-4 Trim front Correct in rear legs	2 Weak top Lacks depth and thickness
Middle pair	2-3 Balance Heavy muscle Smooth shoulder Longer rump Bone	3-2 Deeper middle	3 Coarse shoulder Lacks style
Bottom pair	3-1 Bigger Thick top Thick quarter	1-3 Smoother shoulder	1 Small Narrow front Light muscle

FIGURE 3–5 Be sure to take a small notebook to take notes on the classes. *(Courtesy of National FFA Organization.)*

FIGURE 3–6 Make sure to use the proper terms when describing the class. Use different terms with breeding animals than you do with market animals. *(Copyright 2009 iStockphoto/kyc studio.)*

you present your rationale for the placing it should always be done in comparative terms. Instead of saying the number 2 animal had "heavy muscling," say the number 2 animal was "heavier muscled" than the number 3 animal. Also use market terms when comparing market animals and use breeding animal terms when comparing breeding animals in a class, figure 3-6. For example, when comparing the amount of fat on market animals use the term *finish*. When describing the fat on a breeding animal use the term *cover*. Appendix 1 lists many of the proper terms you can use when comparing animals.

In your presentation remember that you are comparing animals and they should be compared by pairs. Begin with a justification of your top pair. Begin by stating the name of the class and how you placed the class. For example, "I placed this class of crossbred market steers 3-2-1-4." Next explain why you placed the top pair as you did: "In the top pair I placed the number 3 steer over the 2 steer because number 3 was heavier muscled, more structurally correct, and carried a higher degree of finish than the 2 steer." Follow the comparison with a few details about the observation, such as, "The 3 steer was wider down the top, thicker through the round, and carried more muscling through the shoulders than the 2 steer." Next relate any characteristics of the second place animal that was better than the top place animal. This is called a grant. "I grant that the number 2 steer was a longer bodied, deeper sided steer but lacked the overall muscling and finish of the number 3 steer." At this time begin the comparison of the middle placed animals. "In my middle pair I placed the number 2 steer over the number 1 steer because…." Finish up by comparing and justifying your bottom pair. Remember that you only make three comparisons: the top pair, the middle pair, and the bottom pair. Appendix 2 provides sample reasons.

SCORING ORAL REASONS

The official judge who hears your reasons will give you a score that will be added into your overall score. The top score is 50 points and the judge will deduct points as deemed necessary. Usually the judge will be professional and fair. However, keep in mind that you are making an impression on the judge and that impression will translate into your score for the oral reasons. The following are some of the factors that are used to determine your score.

Accuracy

Keep in mind that the most important aspect of the oral reasons in **accuracy**, figure 3-7. The entire point of the reasons is to convince the judge you accurately evaluated the class and can explain why you made the decisions you did. Make sure the points you make are what you actually saw rather than making your presentation sound good. Practice keeping a mental image of the animals as you present your reasons. Remembering details such as a black steer with a white spot or a red pig

FIGURE 3–7 Accuracy is the most important aspect of giving oral reasons. *(Courtesy of National FFA Organization.)*

with black spots can help you not only remember the class but also convince the judge that you remember the class. Work these terms into your reasons. For example, "I placed number 3, the spotted red barrow, over number 2 because . . . ."

Using the proper terms to describe what you observed is a crucial part of being accurate. Putting in terminology that sounds good but is not accurate will result in a low score. Make sure you are using terms that go with the class. As mentioned, you will never be able to be competitive until you can use the proper terminology. This is why it is so important that you memorize the terms and body parts of all the animals in Appendix 1. Be careful not to use terms that describe market animals with a class of breeding animals. Make sure you describe structural defects properly and do not use sheep or hog terms with a class of cattle. For example, sheep have hind saddles, pigs have hams, and cattle have hind quarters. Be careful to use the terms appropriately. If not, the judge will most likely get the impression you do not know how to judge.

Your Appearance

Even though the way you dress may have nothing to do with proper placings and proper evaluations, your **appearance** says a lot about you. Keep in mind that you are making an impression and how you dress makes a statement, figure 3-8. Make sure that you are neatly dressed in official dress (if this is an FFA event) and do not wear a hat or chew gum. If you do not have official dress, wear clean, neat clothes that are not too flashy. Wearing jeans with a t-shirt that has a logo or slogan projects an image of sloppiness. You want to give the impression that you take judging seriously and care enough about the event to dress properly.

Your Presentation

When it is your turn to come before the judge, leave your notes outside or place the notebook in your belt behind you. Promptly enter the room and stop about 6 to 8 feet from the judge, stand up straight, and look the judge in the eye. A good rule to use is to look at the distance between you and the judge. If you were to fall forward and land on the judge, you are too close. Back up a step and relax. Do not appear to be like a soldier at attention, because this conveys the notion that you are nervous and not very confident. Stand with your feet spread about as wide as your shoulders and do not lean over. The idea is to appear relaxed and normal, figure 3-9. The judge may offer you your card but it is generally considered good strategy not to look at or rely on your card. It is permissible to place your arms behind you as long as you can look relaxed. You may keep your arms at your side. Be careful about too much gesturing because such body language often appears artificial as though you are acting.

Use a voice that is only slightly above conversational level. Do not shout or speak too loudly, but talk loud enough to be heard. You will have points deducted if the judge cannot hear or understand what you are saying. A common mistake for beginners is to talk too fast. Keep in mind that the judge is trying to follow what you are saying. Always use proper grammar and make complete sentences that follow a logical sequence. Clearly enunciate your words in a manner that can be easily understood. Always keep in mind that you are being judged on your ability to communicate.

One of the most important factors in presentation is to be confident. Do not try to be eloquent like you are giving a speech, but be sincere and communicate

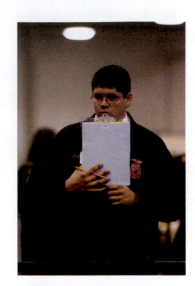

FIGURE 3–8 Keep in mind that you are making an impression and how you dress makes a statement. *(Courtesy of National FFA Organization.)*

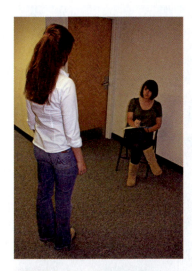

FIGURE 3–9 Stand about 6 feet away from the judge. Make sure your posture is erect, relaxed, and normal.

your justifications in the two-minute limit. Also, avoid being cocky and arrogant. Judges like sincerity, seriousness, and genuineness. The best presentation is emphatic, sincere, and precise. Be polite at all times and when you are finished, thank the judge and ask if there are any questions. A good way to perfect your presentation is to practice before a video camera and play it back. Look for ways of improving your demeanor. Your classmates and teacher can also be of help in critiquing, which will help you confine your presentation to the two-minute limit.

SUMMARY

Most good livestock judges win contests based on the oral reasons. Even though the contestant may differ with the judge on the placing of a close pair of animals, the points lost can be gained back by providing a good set of reasons for the placing. One of the keys to effective oral reasons is a thorough knowledge of the proper terms and their usage. Contestants must have a well-organized, succinct set of reasons to be effective. They must also be able to clearly communicate the rationale for the placings and give the impression of confidence. Good manners and a sharp appearance are vital to obtaining top points in giving oral reasons.

 STUDENT LEARNING ACTIVITIES

1. Judge a class of livestock with your team. Make notes on your placings and trade notes with a team member. Compare how each other organized the notes. Everyone should be able to understand the notes taken by everyone else.
2. Using your notes from activity 1, organize and present a set of oral reasons to your teammates and have them critique you. Incorporate their suggestions and present the reasons again.
3. Set up a video camera and tape yourself giving a set of oral reasons. List ways that you can improve your presentation. Keep doing this exercise until you have perfected your presentation.

 FILL IN THE BLANKS

1. Every science has its own _____ with _____ that are unique to that discipline.
2. Perhaps the most important part of competitive judging is the ability to _____ your _____ and to present your _____ for why you placed the class as you did.
3. You will be given _____ _____ to explain to the judges your rationale for your placing.
4. When you present your rationale for the placing it should be done using _____ terms.
5. Keep in mind that the most important aspect of oral reasons is _____.
6. Practice keeping a _____ _____ of the animals as you present your reasons.
7. Keep in mind that you are making an _____ and how you dress makes a _____.
8. Do not appear to be like a soldier at attention, because this conveys the notion that you are _____ and not very _____.
9. Use a voice that is slightly above _____ _____.
10. The best presentation is _____, _____, and _____.

MULTIPLE CHOICE

1. Judging means that you are:
   a) correct
   b) rendering an opinion
   c) guessing
   d) critical

2. Oral reasons give you the opportunity to:
   a) state the obvious
   b) be creative
   c) justify your placings
   d) none of the above

3. The time limit for presenting oral reasons is:
   a) 2 minutes
   b) 5 minutes
   c) no time limit
   d) 1 minute

4. A good approach is to:
   a) compare the top animal to the bottom animal
   b) compare the top animal with the third place animal
   c) begin with a description of the bottom animal
   d) none of the above

5. The top score for a reasons class is:
   a) 60 points
   b) 100 points
   c) 50 points
   d) 25 points

6. The most important aspect of presentation of oral reasons is:
   a) accuracy
   b) use of sophisticated terms
   c) dress
   d) manners

7. For the oral presentation, you should:
   a) wear a cowboy hat
   b) chew gum
   c) wear a t-shirt with a livestock slogan
   d) none of the above

8. When you stand before the judge, you should:
   a) lean over
   b) stand ramrod straight
   c) appear normal and relaxed
   d) stand close to the judge

9. During the presentation you should:
   a) look the judge in the eye
   b) look at the wall
   c) look down
   d) close your eyes

10. You should talk:
    a) very slowly
    b) loudly
    c) rapidly
    d) clearly and distinctly

DISCUSSION

1. Explain why it is important to learn the proper terminology.
2. Discuss why oral reasons are the most important part of the competition.
3. Describe a good system for taking notes.
4. What is the proper format and organization for presenting reasons?
5. Explain why it is important to talk using comparative terms.
6. What is a grant?
7. Name three main areas used by the judge for scoring oral reasons.
8. Describe how you should dress for presenting oral reasons.
9. What is the proper way to stand?
10. What is the proper way to control your voice?

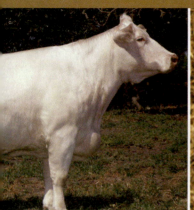

SWINE

CHAPTER 4
Selecting Breeding Swine

OBJECTIVES

As a result of studying this chapter, students should be able to:

- Explain the importance of selecting breeding stock
- Discuss why structural soundness is important in selecting breeding stock
- List the feet and leg problems that can be encountered with structurally incorrect pigs
- Distinguish between a structurally correct pig and one with structural problems
- Identify problems that can be encountered with under lines in gilts
- Distinguish between a reproductively sound breeding animal and one that is unsound

ASPECTS OF SELECTION

As you begin to learn about the selection and evaluation of pigs, keep in mind that there are several aspects of selection. The first aspect that comes to mind is the show arena where animals are showcased for their desirability as breeding or marketing animals. Another aspect is that of the producer who is more interested in profit than in winning shows. Although ideally the type of animals chosen in the show ring should be the same type as those chosen by the producer, this is not always the case. Producers select animals that fit the need of their production program. In the twenty-first century, methods of selecting pigs are much different than in times past. At one time, visual evaluation was the only method the producer or the show ring judge could use in selecting swine, figure 4-1.

Over the years a large body of scientific research has produced a lot of data that can be used in the selection process. Evaluators can now rely on research to determine which characteristics are most likely to be passed from one generation to the next. In the next chapters on selecting swine you will learn how to visually select pigs and also how to use data such as production records in the selection process.

The most important aspect of swine selection is that of evaluating and selecting breeding animals. Keep in mind that the characteristics of the parents will be passed on to the young they produce. This means that an important aspect of selecting animals to use in a breeding program is to select sires and dams that have the characteristics wanted in market animals. Also keep in mind that in order for the producers to make a profit, animals must reproduce efficiently. **Reproductive efficiency** means that offspring must be produced in a predictable, regular basis and the young must be born alive and healthy. In the case of pigs, large litters are important in order to produce the number of piglets needed to be profitable, figure 4-2.

Remember that a litter of ten piglets can be as easily cared for as a litter of four. Obviously ten live, healthy piglets can be much more profitable than four. In addition, the females should be able to deliver her piglets without help from a caretaker. An animal that can give birth on her own is much more efficient than one who needs the help of a human assistant.

Reproductive efficiency is affected by many factors. As this chapter points out, the comfort of the animals depends a lot on how the body of the animal is structured. However, even structurally correct animals can be inefficient if they are reproductively unsound. All these factors must be considered in selecting animals to reproduce.

FIGURE 4–1 Although ideally the type animals chosen in the show ring should be the same type as chosen by the producer, this is not always the case. *(Courtesy of Dr. Frank Flanders.)*

FIGURE 4–2 Perhaps the most important aspect of the pork industry is the ability of sows to produce and wean large litters of pigs. *(© Nemanja Glumac, 2007. Used under license from iStockphoto.com.)*

FIGURE 4–3 Pigs should be selectively bred to live comfortably on hard surfaces such as concrete. *(Courtesy of USDA.)*

FIGURE 4–4 Sows must be able to stand in farrowing crates without undue stress to her joints. *(© 2009 iStockphoto.com/ Edd Westmacott.)*

EVALUATING STRUCTURAL SOUNDNESS IN PIGS

An important aspect of judging or evaluating pigs is structural soundness. **Structural soundness** refers to the skeletal system and how well the bones support the animal's body. Keep in mind that the bones making up the skeletal system are the frame upon which the muscles and internal organs are suspended. The bones must support all the weight of the animal. Bone growth, size, and shape can have quite an effect on the well-being of the animal. Animals that are structured well are more comfortable as they move about or stand in one place. Structural soundness may affect reproductive efficiency. Boars that have problems moving freely are less likely to be interested in breeding than boars that move freely and are comfortable. Also, boars that stand too straight on their legs will have problems mounting females in the mating process. Most of today's hogs are raised on concrete because this surface is much easier to clean and is more sanitary than wood or dirt floors. Pigs should be selectively bred to be comfortable living on the hard surface, figure 4-3.

Structurally sound pigs are more comfortable as they stand and walk on concrete. Structural defects are amplified by the effect of standing and walking on concrete. Soreness, stiffness, and pain in moving greatly reduce reproductive ability. In addition, pigs that have problems standing and moving on concrete generally do not live as long and are not as productive as structurally correct pigs that are more comfortable on concrete.

Most pigs are born in farrowing stalls that protect the young piglets from being crushed as the sow lays down. This means that the female must remain in the crate for a few weeks until the piglets are of sufficient size to wean. The sow must be able to stand up in the crate without undue stress on her joints, figure 4-4.

STRUCTURALLY CORRECT TOP LINES

A skeletal structure that allows a pig to be comfortable on concrete will have a top line that is almost level. The **top line** refers to the length of the top of the animal's back. This can be observed as the outline of the pig as it is standing or moving.

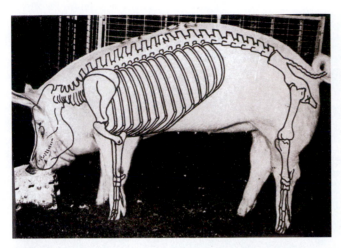

FIGURE 4–5 The scapula and humerus (shoulder and front leg bone) are more vertical and provide less flex and cushion in a pig with a high arch. *(Courtesy of* National Hog Farmer.*)*

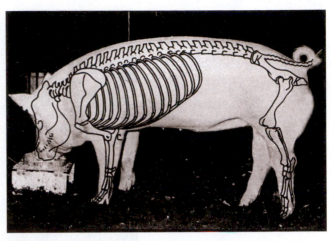

FIGURE 4–6 The same bones are more nearly parallel to the ground in a pig with a level top line. *(Courtesy of* National Hog Farmer.*)*

At one time, pigs were selected for a uniform arch down the top line; however, pigs with a strongly arched back usually have steep rumps and straight shoulders. These characteristics can cause problems for animals that live on concrete. A more modern type animal has a level top that is uniform from top to bottom.

Think of the joints where the legs are attached to the backbone as being hinges that allow the legs to flex and move. The arched topped pig has less flexibility and cushion to the joints. As figure 4-5 shows, the **scapula** and **humerus** (shoulder and front leg bone) are more vertical and provide less flex and cushion than the level topped, more structurally correct pig in figure 4-6. Note the vertical position of the **aitch bone** and **femur** (hip and rear leg bones) of the pig in figure 4-5, as opposed to the same bones that are more nearly parallel to the ground in the pig in figure 4-5.

Again, bones parallel to the ground add more cushion and flex as the animal walks. If these bones are more nearly vertical to the floor, the ends of the bones will jar when the animal walks. This will eventually cause discomfort to the animal. On the other hand, if these bones are closer to being parallel with the floor, they will act more as a hinge. These animals will be more comfortable walking on concrete. The shock of walking will be more absorbed by the **ligaments** of the joints and will have less of a jarring effect to the bones. Ligaments are the structures that connect the bones together.

As mentioned earlier, the top line should be level and uniform. Avoid pigs that have a drop behind the shoulders such as the gilt in figure 4-7. This condition is indicative of a weaker spine and overall weaker structure that may not stand up to walking constantly on concrete. Figure 4-8 shows a gilt with an undesirable high-arched back. Figure 4-9 shows a gilt with a proper level top line.

PASTERN STRUCTURE

FIGURE 4–7 Avoid pigs that have a drop behind the shoulders such as this gilt. *(Courtesy of* National Hog Farmer.*)*

Pasterns (ankle bones) that are too vertical cause too much of a jarring effect to the skeletal system as the animal walks. On the other hand, the pasterns should not be so sloping that they are weak. Notice the steep slope of the pasterns on

FIGURE 4–8 An example of a gilt with an undesirable, arched back.

FIGURE 4–9 A gilt with a proper level top line. *(Courtesy of* National Hog Farmer.*)*

the front and rear legs of the animals in figures 4-10 and 4-11. Weak pasterns can cause pain to the pig as it stands and walks on concrete and can eventually cause the animal to have difficulty walking. In severe cases, the pasterns may slope so much that the **dew claws** can be damaged as they come in contact with the floor. This can be an additional cause of pain for the animal. The correct structure of the pasterns is illustrated in figures 4-12 and 4-13.

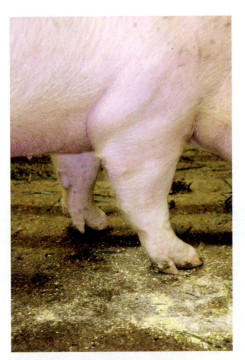

FIGURE 4–10 The pasterns on this gilt have too much slope. Note how they almost touch the ground. *(Courtesy of* National Hog Farmer.*)*

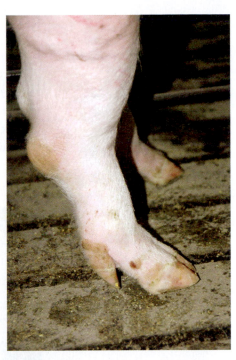

FIGURE 4–11 An example of rear pasterns with too much slope. *(Courtesy of* National Hog Farmer.*)*

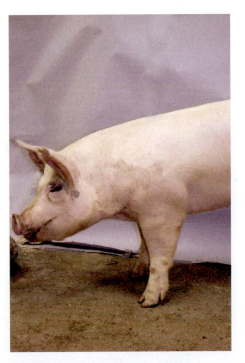

FIGURE 4–12 This gilt has correctly structured front pasterns. *(Courtesy of* National Hog Farmer.*)*

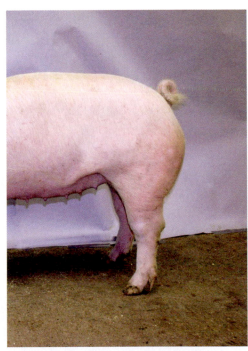

FIGURE 4–13 Correctly structured rear pasterns. *(Courtesy of* National Hog Farmer.*)*

BONE DIAMETER AND FOOT STRUCTURE

Bones should be large in diameter, not small and refined. Larger diameter bones are stronger, and research has shown that animals with larger diameter bones will tend to grow faster. It also stands to reason the larger boned pigs usually have more capacity for muscling. Stronger bones allow the animal to stand and walk more comfortably because there is more volume of bone to carry the weight of the animal. Figure 4-14 shows an animal that has a light, small bone structure. Figure 4-15 depicts an animal with larger and more desirable bone structure.

FIGURE 4–14 The bone structure on this animal is too light. Note the small diameter of the leg bones. *(Courtesy of Dr. Frank Flanders.)*

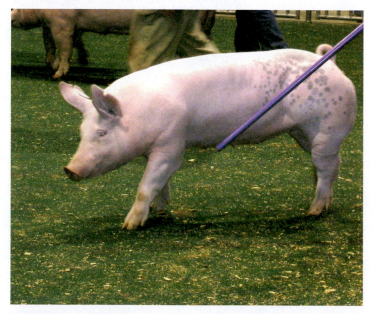

FIGURE 4–15 An animal with sufficient bone size. *(Courtesy of Dr. Frank Flanders.)*

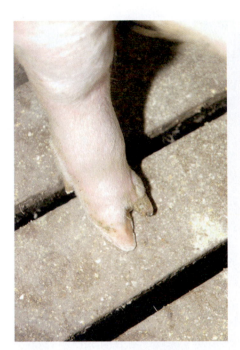

FIGURE 4-16 Note the small inside toe. *(Courtesy of* National Hog Farmer.*)*

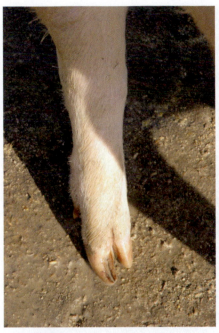

FIGURE 4-17 Another example of a small inside toe. *(Courtesy of* National Hog Farmer.*)*

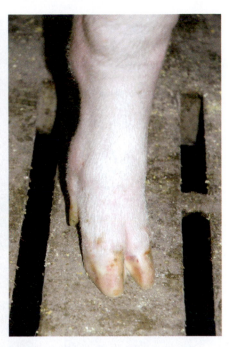

FIGURE 4-18 This animal displays proper toe size. Also, note how large the pad is on the foot. *(Courtesy of* National Hog Farmer.*)*

The toes of the animal's feet should be approximately the same size. Toes that are uneven (most commonly a small inside toe) indicate structural unsoundness, figures 4-16 and 4-17. This inherited defect causes misalignment of the feet, weakens the pasterns, and causes an abnormal amount of weight to be placed on the outside toes. This understandably can cause a great deal of discomfort to the animal. Legs both front and rear should be squarely placed under the animal, figure 4-18.

FEET AND LEG POSITION

The entire legs of the animal must be sound. This means that from the shoulder in the front and the aitch bone in the rear, the legs must be correctly structured. Front feet that are turned out (**splayfooted**) or turned in (**pigeon-toed**) should be avoided. Figure 4-19 shows an animal that is splayfooted. Notice how the front feet point in opposite directions. The condition of the rear feet, cow hocked, shows the same problem—feet that are turned to the outside, figure 4-20.

Pigeon-toed feet also cause problems. Here, the feet also go in opposite directions, but they are turned inward instead of outward. Common sense tells you that animals that walk and stand with feet going in opposite directions cannot stand and walk comfortably. Another undesirable characteristic is a condition called **buck kneed**. This means that the animal's knees are bowed or bucked out in front. Figure 4-21 shows a buck kneed animal. Because this condition does not allow the legs to be placed squarely and straight beneath the animal, joint wear and pain are more likely.

FIGURE 4–19 The front feet of this gilt turns outward. *(Courtesy of National Hog Farmer.)*

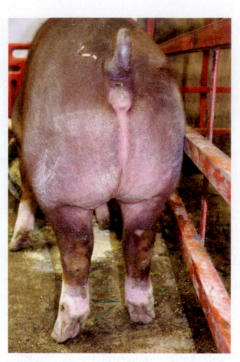

FIGURE 4–20 This pig is cow hocked. Note how the rear feet point outward. *(Courtesy of National Hog Farmer.)*

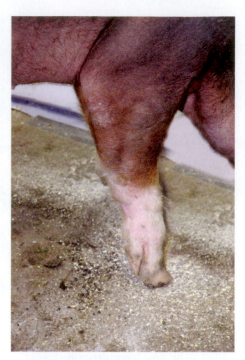

FIGURE 4–21 Notice how the knees of this animal are bowed toward the front. This condition is known as buck kneed. *(Courtesy of National Hog Farmer.)*

FIGURE 4–22 This pig has legs that are structurally correct. Note how the feet and legs are placed squarely under the animal. *(Courtesy of National Hog Farmer.)*

The feet must be straight and in line with the body of the animal. This can be determined as you observe the animal both standing and walking. Figure 4-22 illustrates a gilt with properly placed feet and legs. Note how the legs are squarely under the animal. The gilt appears to be comfortable while standing.

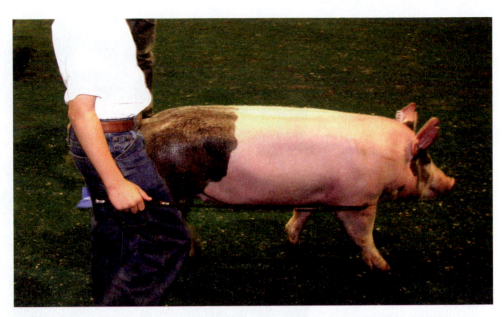

FIGURE 4–23 This pig moves with a long, easy stride. Notice how the rear feet are placed where the front feet were. *(Courtesy of* National Hog Farmer.*)*

The pig should move out with a long, easy stride. As the animal walks, the rear foot should be placed about where the front feet were placed. Pigs that take short, choppy steps (goose stepping) are either structurally unsound, muscle bound, or both. Structurally correct animals move with a fluidlike grace that looks comfortable as opposed to animals that look like they are struggling as they walk, figure 4-23.

CAPACITY

Capacity refers to the width and depth of an animal, or its total volume. Preference should be given to those animals that are wide down the top and deep in the side. The reasoning is that those animals that have greater dimensions in the side, down the top, and through the chest have more room for the vital organs such as the heart and lungs. In addition, pigs with wider belly have a greater capacity for holding feed and thus gain more rapidly. The ribs should be long and well arched. The rib cage should be rectangular; that is, the fore rib should be about as long as the rear rib, figure 4-24.

Breeding animals should be long down the side. Remember that the dam and sire pass along their characteristics to their offspring. It stands to reason that longer bodied pigs tend to yield more meat than shorter bodied pigs. Also keep in mind that a longer bodied gilt has more room for a larger number of well-placed teats.

Pigs should stand wide with adequate distance between the legs. The animal should be wide between both the hind and the front legs and have a wide chest floor. Animals that stand with feet close together are usually narrower throughout the body cavity. Also closely placed feet and legs indicate lack of muscling. Animals should stand wide and square, as shown in figure 4-25. The gilt in figure 4-26 stands narrow and lacks both body capacity and muscling.

FIGURE 4–24 This animal has a long, square rib cage that provides room for the internal organs. *(Courtesy of* National Hog Farmer.*)*

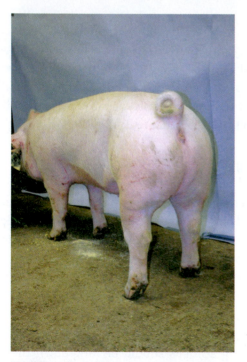

FIGURE 4–25 This pig stands wide and square. *(Courtesy of* National Hog Farmer.*)*

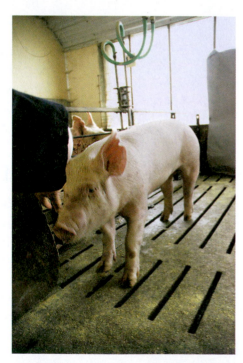

FIGURE 4–26 Notice how narrow this gilt is. She lacks both body capacity and muscling.

REPRODUCTIVE SOUNDNESS

As mentioned, the factors in swine that add profitability to the raising of pigs are reproductive efficiency, growthiness, and carcass merit. Reproductive efficiency and growthiness are by far the two most important. With these factors in mind, breeding hogs are evaluated for characteristics that best combine these factors. To be reproductively efficient, a female must be feminine; that is, she must look like a female. The same substances (**hormones**) that control the reproductive cycle

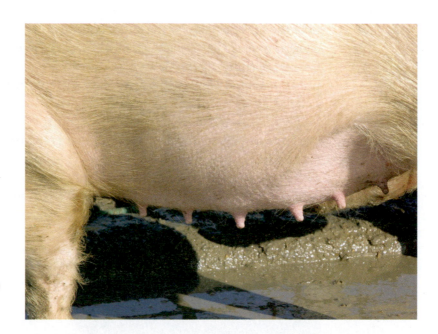

FIGURE 4–27 A female should have at least six pairs of evenly spaced teats. This is an example of a good under line. *(Courtesy of* National Hog Farmer.*)*

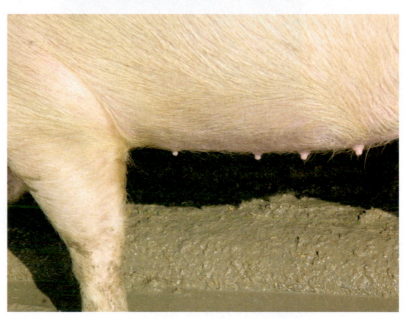

FIGURE 4–28 The teats on this female are too closely spaced. *(Courtesy of* National Hog Farmer.*)*

also account for the development of sex characteristics. The sex characteristics are therefore an indication that the female is producing a large enough quantity of hormones to cause her to conceive efficiently.

There are several indicators of femininity that a producer can use to select females that are reproductively efficient. The under line should be well defined; that is, the teats should be large and easily seen. Remember that the mammary glands produce milk for the piglets and a malformed mammary or teat cannot produce the milk necessary for the piglets to grow and remain healthy. A strong factor in the growth rate of the piglets is the amount of milk they receive from the mother. There should be at least six pairs of prominent and evenly spaced teats, figure 4-27. If the teats are too close together, there may not be enough surrounding mammary tissue for good milk production, figure 4-28. Pin, blind, or inverted nipples should be avoided. **Pin nipples** are very tiny nipples that are much smaller than the other nipples on the under line, figure 4-29. These may not function well enough to feed the young pigs. **Blind nipples** are nipples that fail to mature and have no opening, figure 4-30. Obviously these have little use as they are nonfunctional. **Inverted nipples** appear to have a crater in the center and are not functional. Usually a female that is producing enough female hormones will have the proper under line to efficiently feed the young she bears. Gilts and sows should look like females—they should have a head that is shaped like a female not like a boar, figure 4-31. Although all pigs selected for breeding should have a large, broad head, gilts should not have the massive head that is characteristic of the boar.

Another defect that adversely affects reproduction is an **infantile vulva**, figure 4-32. This condition is characterized by a very tiny vulva in a breeding age

FIGURE 4-29 Pin nipples are tiny nipples that are much smaller than the other nipples. *(Courtesy of National Hog Farmer.)*

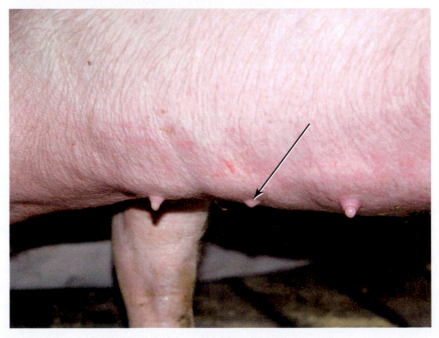

FIGURE 4-30 Blind nipples have no opening and supply no milk. *(Courtesy of National Hog Farmer.)*

FIGURE 4–31 The head of a boar should look rugged and masculine. *(Courtesy of ARS.)*

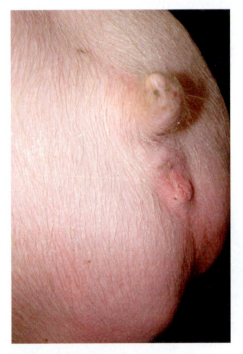

FIGURE 4–32 This gilt has an infantile vulva that is underdeveloped. *(Courtesy of National Hog Farmer.)*

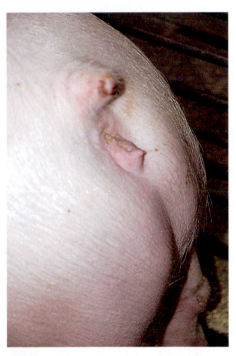

FIGURE 4–33 The tipped vulva on this gilt may lead to breeding problems. *(Courtesy of National Hog Farmer.)*

gilt. This makes breeding difficult and conception rates are usually very poor. If the vulva is small, then this is a good indication that the entire reproductive tract is small. Even if the gilt conceives, she may have great difficulty delivering the piglets. A gilt with an infantile vulva should be culled from the herd. Another defect is a vulva that is tilted or tipped upward, figure 4-33. This condition can make the

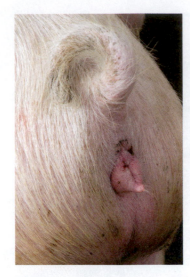

FIGURE 4–34 Example of a normally formed vulva. *(Delmar/Cengage Learning, Photo by Ray Herren.)*

act of mating difficult. Gilts should be selected that have large, normally formed vulvas, figure 4-34.

Boars should appear massive, rugged, and masculine. The **testicles** should be large and well developed inside a scrotum that is well attached. Research has shown that the larger the male's testicles, the more **viable** sperm he will produce and the more aggressive he will be in breeding, figure 4-35. Large, pendulous, or swollen **sheaths** should be avoided as these characteristics can lead to breeding problems.

MUSCLING

Breeding animals must have adequate muscling to pass the trait along to their offspring. However, if the gilt or boar has too much muscling, problems can occur. For example, if a gilt has tightly wound muscling in the ham region, there may not be adequate room for the reproductive tract. This may cause problems not only when attempting to conceive, but also when giving birth as the muscling may constrict the birth canal, figure 4-36. Gilts should have smooth muscling that does not appear to bulge, figure 4-37. Boars with too much tight muscle may have problems mounting and attempting to breed. Excess muscling can also restrict the proper movement of the animals. The proper structure for muscling will be explained in the next chapter on selecting market pigs.

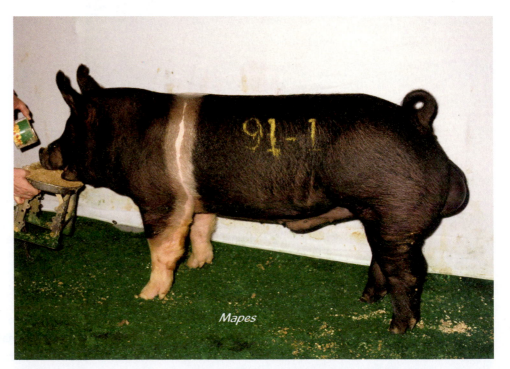

FIGURE 4–35 Boars should have large testicles that are well attached. *(Courtesy of Heimer Hampshires.)*

FIGURE 4–36 This gilt shows round bulging muscles in the hams, which may lead to breeding and farrowing problems. *(Courtesy of Dr. Frank Flanders.)*

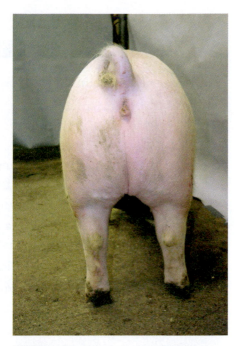

FIGURE 4–37 Note the smooth muscling on this gilt. She has adequate muscling but the muscling is shaped correctly. *(Courtesy of* National Hog Farmer.*)*

SUMMARY

Visual evaluation of breeding swine is an important task for producers. The characteristics of the parent stock will be passed on to their offspring that are to be raised for the market. Selecting pigs that are structurally sound will help ensure that the animals can move and live comfortably in a confinement operation. Pigs that are more comfortable are more productive. Breeding soundness is just as important. If the animals cannot breed and produce young efficiently, they will be of little use.

Visual evaluation is a major part of selecting breeding swine, but by no means the only factor to be used. In the next chapter, you will learn how to interpret productions records and data to be used in the selection process.

STUDENT LEARNING ACTIVITIES

1. Conduct a search of the internet for images of pigs. Try to find examples of pigs that are not as structurally correct as desired by producers. Identify the structural defect by its proper name. Compare your images with others in the class.
2. Contact producers in your area and ask for any advertising images of their breeding stock. If there are none in your area, addresses of producers can be found on the internet.

FILL IN THE BLANKS

1. The most important aspect of swine selection is evaluating and selecting _____ _____.
2. Bone _____, _____, and _____ can have quite an effect on the well-being of the animal.
3. Pigs with a strongly _____ _____ usually have steep rumps and straight shoulders.
4. Boars that have problems moving freely are less likely to be interested in _____ than boars that move _____ and are _____.
5. _____ are the structures that connect the bones together.
6. In severe cases, the pasterns may slope so much that the _____ _____ can be damaged as they come in contact with the _____.
7. _____ refers to the width and depth of an animal.
8. The factors in swine that add profitability to the raising of pigs are _____ _____, _____, and _____ _____.
9. _____ _____ are very tiny nipples that are much smaller than the other nipples on the under line.
10. Boars should appear _____, _____, and _____.

MULTIPLE CHOICE

1. Large liters are important because:
   a) they can be as easily cared for as small liters
   b) they are more profitable
   c) more pigs can be produced
   d) of all of the above
2. Structurally correct animals:
   a) look better
   b) are more productive
   c) are less comfortable on concrete
   d) take smaller steps
3. Structural correctness:
   a) does not affect growth
   b) can have a great effect on reproductive efficiency
   c) is of minor concern to producers
   d) can cause soreness in joints
4. Most pigs are born:
   a) in farrowing crates
   b) on straw floors
   c) in the open
   d) on finishing floors
5. The top line should:
   a) be uniformly arched
   b) have a dip behind the shoulders
   c) be level
   d) be slightly arched
6. The aitch bone and femur should be:
   a) vertical to the floor
   b) parallel to the floor
   c) close to the floor
   d) square to the floor
7. Ligaments are structures that connect:
   a) muscles
   b) fat to muscle
   c) bones
   d) skin to muscle
8. Bones should be:
   a) small and refined
   b) as short as possible
   c) all the same size
   d) large in diameter

9. Which of the following is not a term used to describe a mammary problem?
   a) blind nipple
   b) pin nipple
   c) prominent nipples
   d) inverted nipple

10. A small vulva can cause which of the following problems?
    a) poor conception
    b) trouble in breeding
    c) difficult birth
    d) all of the above

 DISCUSSION

1. What is meant by reproductive efficiency?
2. Why are large litters important?
3. Explain why animals need to be structurally sound.
4. Why is a level top line so important?
5. Why are pasterns that are too vertical a problem in breeding animals?
6. Why should the toes of the animal be the same size?
7. List the terms that describe leg defects.
8. Why is it important that an animal have a wide chest and wide ribs?
9. What role do hormones play in the appearance of a breeding animal?
10. Describe how a sound under line should look on a gilt.

CHAPTER 5

The Selection of Market Swine

OBJECTIVES

As a result of studying this chapter, students should be able to:

- Discuss some of the trends in selecting market hogs over the past decades
- Outline some of the problems encountered with the types of pigs deemed as ideal during the decades since the 1950s
- Describe the ideal modern market pig
- Explain how to visually distinguish between a fat and a lean pig
- Explain how to distinguish between a thick muscled pig and a thinly muscled pig
- Describe a pig that has too much muscling
- Explain why structural soundness is important in selecting market pigs
- Explain how live market pigs are graded

MARKET PIG TRENDS

Pigs have been raised for food as far back as 4900 BC. Ancient records show that pigs were domesticated and raised in China. They were introduced in America by Hernando Desoto in 1539. Since then, the pig has been an important part

FIGURE 5-1 At one time almost all farms produced pigs. *(Courtesy of Nebraska State Historical Society.)*

of American agriculture. Not many years ago almost all American farms produced pigs as part of the farm income or for food for the table, figure 5-1. As with all domesticated animals, pigs have gone through many years of selection, and over the years there have been many changes in the way pigs are selected. The type of animal produced has undergone a lot of change as the needs of the industry have changed. However, with all the modifications in what is determined as the perfect type of pig, several factors remain constant. First, the producer must make a profit. If a profit cannot be made there will be very few pigs produced for market. Although market factors come into play, profit is often determined by reproductive efficiency, growth ability, and feed conversion of the animals.

As pointed out in Chapter 4, reproductive efficiency refers to the ability of animals to give birth on a regular basis, have a large litter, and provide enough milk to get the piglets off to a good start. Even if pigs reproduce efficiently, they must have the ability to grow efficiently on a reasonable amount of feed and produce a carcass that in turn produces an acceptable amount of quality pork. All these factors come into play when selecting both breeding and market animals. Breeding animals must have the potential to produce market animals with the growth and carcass traits that will allow the producer to make a profit.

Another large factor in the selection of market pigs is the needs and desires of the consumer. This factor has influenced the selection and production of pork for many years. Prior to the 1950s, swine were raised primarily for lard. People used the lard not only for frying and cooking food but also in the manufacture of cosmetics and lubricants, figure 5-2. About that time, health concerns over the effects of animal fat such as lard became an issue in human diets. With the advent

MIDDLE-BRED WHITE PIGS.

FIGURE 5–2 Prior to the 1950s, pigs were used primarily for lard production. *(Photo by Hulton Archive/Getty Images. Used under license from iStockphoto.com.)*

of healthier cooking oils made from vegetable oils and cosmetics and lubricants made from petroleum-based synthetics, the demand for lard was dramatically reduced. Efforts were then put into developing a hog that produced meat instead of lard. Pigs that were used for breeding were especially selected for the degree of muscling they possessed. The idea was to produce an animal with the maximum amount of meat and the minimum amount of fat. These efforts culminated in the late 1960s and early 1970s with the so-called super pig. Pigs were selected for huge, round, bulging hams and overall thickness of muscling. Although producers were successful in producing a very lean pig that had a lot of muscle, some problems arose with the highly muscled, extremely lean pigs. This type of pig failed to be the ideal type for three reasons.

1. Porcine stress syndrome. Extremely heavy muscled pigs are associated with a condition known as **porcine stress syndrome (PSS)**. Apparently this condition is genetic and is passed on to the offspring by the parents. Pigs suffering from PSS have very little tolerance to stress associated with hot weather, moving about, and some management practices. When put under such stress, animals with this condition have muscle tremors and twitching, red splotches develop on their underside, and they suffer sudden death, figure 5-3. Obviously this condition in hogs is not in the best interests of the pigs or the producer.

2. PSE pork. Pigs with extremely heavy muscle tend to produce lower quality pork. Although they may have a large quantity of muscling, the meat has little or no intermuscle fat (marbling), is very pale in color, and is soft and watery (exudative). These characteristics account for the name of the condition, **pale, soft, and exudative (PSE) pork**, figure 5-4. Consumers reject this type of pork because the pale color is not appealing and, when cooked, the meat is dry and lacking in taste.

FIGURE 5–3 This pig is suffering from PSS. Note the red splotches on the underline.

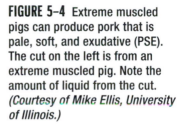

FIGURE 5–4 Extreme muscled pigs can produce pork that is pale, soft, and exudative (PSE). The cut on the left is from an extreme muscled pig. Note the amount of liquid from the cut. *(Courtesy of Mike Ellis, University of Illinois.)*

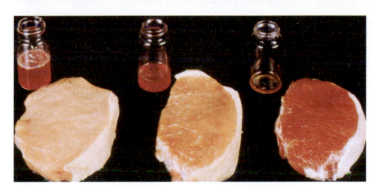

3. Less reproductive efficiency. Heavily muscled, tightly wound boars have problems moving about and mounting females that are in heat. In addition, their sperm count is often very low. These conditions make them less desirable as herd sires. Females that are too heavily muscled are less fertile and have problems conceiving. Those that do become pregnant often have problems farrowing because the birth canal is so tightly bound with muscling that it cannot expand properly. Also the number of piglets born is often fewer than those from less muscular females.

In an effort to correct these problems, during the 1970s producers developed long, tall, flat muscled pigs. The idea was to produce animals that could move freely and reproduce efficiently. Market emphasis was on pigs that were extremely long, tall, and could move freely. Pigs with extreme bulge and flare to their muscle pattern were highly discriminated against. The tall, flat muscled pigs had greater resistance to stress and could reproduce more efficiently. Boars were more efficient breeders and sows had larger litters.

Although this type of pig improved reproductive efficiency, problems were encountered with carcass desirability. The amount of muscle on these animals did

FIGURE 5–5 The loin eye is the cross section of the long muscle that runs down both sides of the backbone. This cut is from the loin eye. *(Courtesy of Dr. Frank Flanders.)*

not suit the demands of the packers. **Loin eye** areas (an indication of the overall amount of muscle in the carcass) began to be unacceptably small. The loin eye is a cross section of the long muscle that runs down both sides of the backbone of the pig, figure 5-5. This is the muscle where retail cuts of pork chops and pork loin are taken. In addition, the growth rate and feed efficiency of these pigs were lower than the producers wanted. Market conditions also influenced the type of pig that was needed. Added production costs demanded that producers raise a more efficient pig.

Beginning in the 1980s, the emphasis on selection in pigs was on what was termed the high-volume pig: very wide down the top, especially at the shoulders. Producers wanted pigs that were very deep and widely sprung at the ribs and deep in the flank and belly, because this type of pig had a lot of room in the body cavity for the internal organs such as the heart and lungs. An animal with larger internal organs seemed to grow faster and remain healthier. Also, in a radical change from years past, pigs were selected that had large, loose bellies that were capable of holding large amounts of feed. These animals were more efficient in their intake of feed and in the conversion of feed to body weight. In addition, these pigs with larger bellies produced more bacon, because bacon comes from the belly portion of the pork carcass. A wide topped, deep sided animal has more capacity and internal volume for the internal organs such as heart, lung, and digestive tract. This type of animal is a "better doing animal" in that it should have a higher rate of gain than a narrow topped, shallow sided animal.

In the 1990s, the trend was toward a leaner, more muscular pig. Emphasis was on leanness and pigs that had little wastage in the carcasses. Some of the same problems were encountered in the 1960s. Consumers complained that the sausages were too dry and some fat was needed to flavor the pork. The proper amount of fat has always been a dilemma. Too much fat is a waste both from the standpoint of the amount of feed needed to put on the fat and the excess that has to be trimmed from the carcass. Too little fat and the pork will be dry and tasteless.

Beginning in the 1980s, the National Pork Board began to develop a conceptualization of what was considered the ideal pig. The effort was a collaboration of producers, educators, and scientists who came to a consensus as to what they thought the ideal pig should look like and also how the pig should perform. They called the pig Symbol. In 1996, the board revised the concept of Symbol and created Symbol II. The current revision was done in 2005 with Symbol III, figure 5-6. Symbol III is not only a visual example, but also an example of the data and research component of the ideal pig. The following are some of the desirable traits outlined by the National Pork Board.

Production Characteristics
- **Live-weight feed efficiency** of 2.4 (2.4)*—This means that the pig should gain 1 pound for every 2.4 pounds of feed consumed.
- **Fat-free lean gain efficiency** of 5.9 (5.8)*—This means that the pig should gain 1 pound of lean meat for every 5.9 pounds (for a barrow) and 5.8 pounds (for a gilt) of feed consumed.
- Fat-free lean gain of 0.95 pound per day
- Marketed at 156 (164) days of age

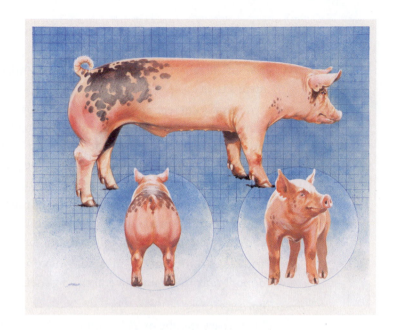

FIGURE 5–6 The National Pork Board developed a conceptualization of a modern type pig. This is Symbol III, created in 2005. *(Courtesy of National Pork Board.)*

- Weighing 270 pounds
- All achieved on a corn-soy equivalent diet from 60 pounds—These are the feeds that have proven to be the most economical for growing out pigs.
- Free of all internal and external parasites
- From a high health production system
- Immune to or free of all economically important swine diseases
- Produced with environmental assurance—This means the pig was raised using approved methods that help protect the environment.
- Produced under **Pork Quality Assurance (PQA™)** and **Transport Quality Assurance (TQA™)** from the National Pork Board. According to the board, the program "emphasizes good management practices in the handling and use of animal health products, and encourages producers to review their approach to their herds' health programs." The guidelines include the regulation of the quality of raw materials, assemblies, products, and components; services related to production; and management, production, and inspection processes. The TQA program "encourages dedication to transporting and delivering the highest quality, safest product possible to remain competitive in U.S. and world markets and is to be the consumer's meat of choice." By completing this program, truckers demonstrate their commitment to "quality assured pork transportation and delivery."
- Produced in a **Swine Welfare Assurance Program (SWAP)**, which is an education and assessment tool designed to promote and continually enhance swine welfare on the farm
- Free of the stress gene (halothane 1843 mutation) and all other genetic mutations that have a detrimental effect on pork quality
- Result of a systematic crossbreeding system, emphasizing a maternal dam line and a terminal sire selected for growth, efficiency, and superior muscle quality

- From a maternal line weaning more than 25 pigs per year after multiple parities
- Free of all abscesses, injection site blemishes, arthritis, bruises, and carcass trim
- Structurally correct and sound with proper angulations and cushion and a phenotypic design perfectly matched to the production environment. The **phenotype** or phenotypic characteristics refer to how the animal appears—its physical characteristics. The **genotype** or genotypic characteristics refer to the actual genetic makeup of the animal.
- Produced in a system that ensures the opportunity for stakeholder profitability from the producer to retailer while providing a cost competitive product retail price in all domestic and export markets
- Produced from genetic lines that have utilized genomic technology to support maximum improvement in genetic profitability and efficiency

Carcass Characteristics
- Hot carcass weight of 205 pounds—The hot carcass weight is the weight of the carcass immediately after it has been killed and dressed.
- LMA of 6.5 (7.1)* square inches—This refers to the area of the cross section of the loin muscle.
- Belly thickness of 1 inch
- Tenth rib back fat of 0.7 (0.6)* inch
- Fat-free lean index of 53 (54.7)*

Quality Characteristics
- Muscle color score of 4.0
- 24-hour **pH** of 5.9. The pH refers to the amount of acidity in the carcass.
- Maximum **drip loss** of 2.5%. Drip loss is the amount of weight loss after a carcass has cooled over a period of time.
- Intramuscular fat level of 3%, figure 5-7

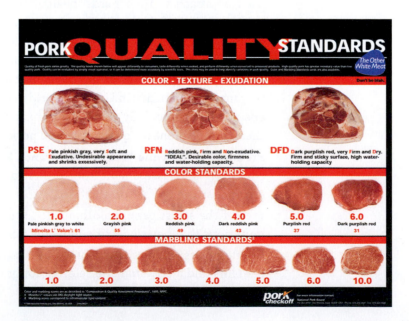

FIGURE 5-7 This image represents the standards for pork as set by the National Pork Board. *(Courtesy of National Pork Board.)*

- Free of within muscle color variation and coarse muscle texture
- Free of ecchymosis (blood splash). This is a splotchy skin condition caused by ruptured blood vessels. It is associated with PSS.
- Provides an optimum balance of nutrients important for human nutrition and health
- Provides a safe, wholesome product free of all violative residues and produced and processed in a system that ensures elimination of all food-borne pathogens

Note: *Numbers in parentheses represent gilt numbers corresponding to the barrow numbers.

DISTINGUISHING MUSCLE FROM FAT

One of the first and most basic skills to learn in judging market hogs is the ability to distinguish between a hog that is thick with muscle and one that is thick because of fat. Obviously a thinly made, narrow pig such as that in figure 5-8 does not have adequate muscle. Notice how the animal stands closely on its legs and how narrow it is down the top. Thicker made pigs are a little more difficult to discern to the untrained eye.

Closely observe the animal from the side, front and back, and over the top. From a side view, a fat, light muscled pig will appear sloppy, particularly at the elbow pocket and base of the ham. These areas will appear soft and will have wrinkles in the skin. The jowls will appear to droop and the hams will be straight down and flat from the tail to the base of the ham. From the back, the legs will be closely set and the hams will show little or no definition of where the muscles join, figure 5-9. The top line will appear to be square from side to side, and it will be difficult to distinguish where the loin meets the ham. Notice the top of the shoulders. If the shoulder blades are not visible, the pig is probably carrying too much fat, figure 5-9.

FIGURE 5–8 A narrow made, light muscled pig. *(© Yui, 2009. Used under license from Shutterstock.com.)*

In contrast, a trim, well muscled pig will appear clean with few wrinkles in the skin. The animal will stand wide, both front and back. Note how Symbol III appears to stand in figure 5-6. The legs are square under the animal and give the appearance of being widely set because of the muscle structure. Looking over the top line, the pig should have a v-shaped line down the backbone. An overly conditioned (too fat) pig will not have this line because it is filled with fat. You should be able to see a definite juncture where the ham and loin join, figure 5-10. The

FIGURE 5–9 This pig is excessively fat. Notice that the widest part of the pig is at the top.

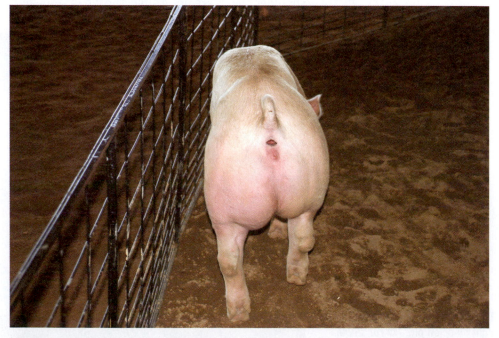

FIGURE 5–10 On a lean, well muscled pig, you should be able to see where the ham and loin join. Notice that the widest part of the ham is through the center.

muscles in the ham should be full and deep with the widest part of the ham being through the lower center portion of the ham. A light muscled, fat pig will be the widest at the top of the hams. A properly muscled and conditioned pig should display distinct muscling in the hams. You should be able to see where the muscles are joined together, figure 5-11. There should be curvature to the hams and not the straight, flat appearance such as that in the fat, light muscled pig. From the front view, the pig should be wide standing and well muscled over the shoulder, figure 5-12.

FIGURE 5–11 Notice the distinct muscles in the ham and shoulder of this pig.

FIGURE 5–12 This pig is well muscled over the shoulders.

FIGURE 5-13 Note the fat deposits in the cushion of the hams and the muscle seams of the gilt on the right as opposed to the leaner gilt on the right. *(© 2009 iStockphoto.com/Eric Isselee.)*

A market pig can be very muscular and still have too much condition (fat). The two gilts in figure 5-13 provide a good example. Notice the lean characteristics of the gilt on the left. The muscles are well defined and the area where the cushion of the ham (where the leg joins the body) is tight and firm. In contrast, the gilt on the right has pockets of fat between the muscles and gives a smoother appearance. Also there are fat deposits in the cushion of the ham. Note that this animal stands as wide as the gilt on the left because she has about the same degree of muscling.

Keep in mind that although the purpose of producing market hogs is to produce muscle, pigs can have too much muscling. This condition can lead to the problems explained earlier in the chapter. Overly muscled pigs have round, bulging hams that give the appearance of basketballs. These muscles protrude from the hams and shoulders and seem unnatural. Note the muscling pattern in the pig in figure 5-14. This pig definitely has too much muscling. Muscle bound pigs have difficulty walking and move in short, choppy steps. A properly muscled pig should walk out with a long, easy stride. The animal should place the hind foot where the front foot was placed as it walks.

STRUCTURAL SOUNDNESS

One line of thought is that pigs headed for slaughter do not necessarily have to be structurally sound. The point is that they will not reproduce and therefore little attention needs to be spent selecting pigs that have feet and legs that are correctly formed and allow the animal to move and stand comfortably. There may be some merit to this thinking, but keep in mind that when you select an animal as desirable in a judging competitive event, you are saying that this is the type of

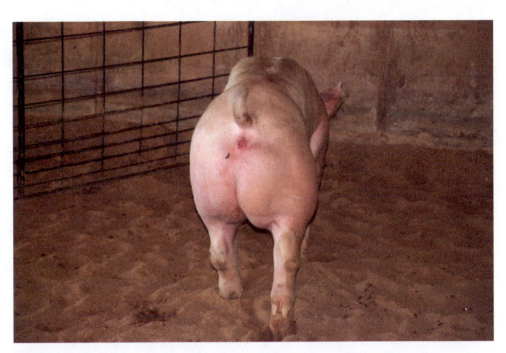

FIGURE 5–14 This pig has too much bulging muscle. Note the round basketball appearance of the hams.

animal we should aim to produce. Remember from the last chapter that structural correctness in pigs is of the utmost importance, even in the selection of market animals. These animals must be comfortable while living on concrete in order to grow and finish efficiently. Review the last chapter on selecting breeding animals. The section on structural (skeletal) soundness applies to market swine as well.

GRADING MARKET HOGS

The U.S. Department of Agriculture (USDA) has set up guidelines for placing grades on pigs that are marketed for slaughter. Because the grade is an indication of the amount and quality of marketable meat that can be expected from the carcass, the grade helps buyers in pricing the pigs. The grades are US No. 1, US No. 2, US No. 3, US No. 4, and US Utility. The grades are based on the prediction of carcass yield and quality. **Carcass yield** is the percentage of lean retail cuts that comes from a carcass. **Quality** refers to the desirability of the pork and includes such measures as firmness of the fat and thickness of the belly fat. A pig that has less than 0.6 inch of belly fat is considered unacceptable. The belly area is the source for bacon, and a belly with fat coverage less than 0.6 inch is considered too thin.

Specifications for Official U.S. Standards for Grades of Slaughter Barrows and Gilts

[The following is taken directly from the USDA standards:]

    (a) The grade of a slaughter barrow or gilt with indications of acceptable quality is determined on the basis of the following equation:

Grade = (4.0 × last rib backfat thickness, inches) (1 × muscling score)

TABLE 5-1 Preliminary Grade Based on Backfat Thickness Over the Last Rib

PRELIMINARY GRADE	BACKFAT THICKNESS RANGE
US No. 1	Less than 1.00 inch
US No. 2	1.00 to 1.24 inches
US No. 3	1.25 to 1.49 inches
US No. 4	1.50 inches and over[1]

[1]*Animals with an estimated last rib backfat thickness of 1.75 inches or over cannot be graded U.S. No. 3, even with thick muscling.*

To apply this equation, muscling should be scored as follows: thin (inferior) = 1, average = 2, and thick (superior) = 3. Animals with thin muscling cannot grade US No. 1. The grade may also be determined by calculating a preliminary grade according to the schedule shown in Table 5-1 and adjusting up or down one grade for superior or inferior muscling, respectively.

(b) The following descriptions provide a guide to the characteristics of slaughter barrows and gilts in each grade.

  (1) US No. 1

    (i) Barrows and gilts in this grade are expected to have an acceptable quality of lean and belly thickness and a high expected yield (60.4% and over) of four lean cuts. US No. 1 barrows and gilts must have less than average estimated back fat thickness over the last rib with average muscling, or average estimated back fat over the last rib coupled with thick muscling, figure 5-15.

    (ii) Barrows and gilts with average muscling may be graded US No. 1 if their estimated back fat thickness over the last rib is less than 1 inch. Animals with thick muscling may be graded US No. 1 if their estimated back fat thickness over the last rib is less than 1.25 inches. Barrows and gilts with thin muscling may not be graded US No. 1.

  (2) US No. 2

    (i) Barrows and gilts in this grade are expected to have an acceptable quality of lean and belly thickness and an average expected yield (57.4% to 60.3%) of four lean cuts. Animals with average estimated back fat thickness over the last rib and average muscling, less than average estimated back fat thickness over the last rib and thin muscling, or greater than average estimated back fat thickness over the last rib and thick muscling will qualify for this grade.

    (ii) Barrows and gilts with average muscling will be graded US No. 2 if their estimated back fat thickness over the last rib is 1 to 1.24 inches. Barrows and gilts with thick muscling will be graded US No. 2 if their estimated back fat thickness over the last rib is 1.25 to 1.49 inches. Barrows and gilts with thin muscling must have less than 1 inch of estimated backfat over the last rib to be graded US No. 2, figure 5-16.

  (3) US No. 3

    (i) Barrows and gilts in this grade are expected to have an acceptable quality of lean and belly thickness and a slightly low expected yield (54.4% to 57.3%) of four lean cuts. Barrows and gilts with average

FIGURE 5–15 A side and rear view of a US No. 1 pig. *(Courtesy of USDA/GDA.)*

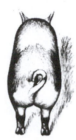

FIGURE 5–16 A US No. 2 pig. *(Courtesy of USDA/GDA.)*

FIGURE 5–17 A US No. 3 pig. *(Courtesy of USDA/GDA.)*

muscling and more than average estimated back fat thickness over the last rib, thin muscling and average estimated backfat thickness over the last rib, or thick muscling and much greater than average estimated back fat thickness over the last rib will qualify for this grade.

(ii) Barrows and gilts with average muscling will be graded US No. 3 if their estimated back fat thickness over the last rib is 1.25 to 1.49 inches. Barrows and gilts with thick muscling will be graded US No. 3 if their estimated back fat thickness over the last rib is 1.5 to 1.74 inches. Barrows and gilts with 1.75 inches or more of estimated back fat thickness over the last rib cannot grade US No. 3. Barrows and gilts with thin muscling will be graded US No. 3 if their estimated back fat thickness over the last rib is 1 to 1.24 inches, figure 5-17.

(4) US No. 4

   (i) Barrows and gilts in this grade are expected to have an acceptable quality of lean and belly thickness and a low expected yield (less than 54.4%) of four lean cuts. Barrows and gilts in the US No. 4 grade always have more than average estimated back fat over the last rib and thick, average, or thin muscling, depending on the degree to which the estimated back fat thickness over the last rib exceeds the average.

   (ii) Barrows and gilts with average muscling will be graded US No. 4 if their estimated back fat thickness over the last rib is 1.5 inches or greater. Barrows and gilts with thick muscling will be graded US No. 4 with estimated back fat thickness over the last rib of 1.75 inches or greater, and those with thin muscling will be graded US No. 4 with 1.25 inches or greater estimated back fat over the last rib, figure 5-18.

(5) US Utility. All barrows and gilts with probable unacceptable quality of lean or belly thickness will be graded US Utility, regardless of their muscling or estimated back fat thickness over the last rib. Also, all barrows and gilts which may produce soft and/or oily fat will be graded US Utility. (Figure 5-19)

FIGURE 5–18 A US No. 4 pig. *(Courtesy of USDA/GDA.)*

FIGURE 5–19 US Utility. *(Courtesy of USDA/GDA.)*

GRADING LIVE ANIMALS

Because grading is based on carcass yield and quality, the grading of live animals is based to a large degree on visual evaluation of the pigs. An exception is the use of ultrasound to determine the amount of back fat on the animal. The device is placed on the back of the animal over the tenth rib and a reading is taken. A sound wave is transmitted that registers against the lean meat in the loin and the distance traveled by the sound wave indicates the depth of the back fat. The procedure is painless to the animal.

The grade you place the animals in will depend on your estimate of the amount of muscling and the depth of back fat you estimate. Go by the USDA rules and guidelines as provided in this chapter.

STUDENT LEARNING ACTIVITIES

1. You have been asked to grade a market pig. As you observe the barrow from all angles, both standing and walking, you determine that the pig has average muscling. You notice that you cannot see the shoulder blades as the pig walks and he appears to have loose jowls and loose wrinkled skin around the bottom of the ham. From this you judge the pig to have about 1 inch of back fat. What grade do you give the animal?

2. You observe a 250 pound gilt that has hams that are the widest at the center. From a rear view, you observe that the muscles can be seen and they are long and well formed from top to bottom. The loin is wide with a distinctly observable juncture. From these observations you conclude that the pig has thick muscling. The skin of the animal is tight and you can clearly see muscle movement as the animal walks. The shoulder blades are visible. You conclude that the animal has about 0.6 inch of back fat. What USDA grade do you assign to this animal?

3. Go to the internet and look up the price of market pigs. What is the price difference between the different grades of pigs?

4. Visit a swine production farm and judge several classes of market pigs. Give a market grade to each animal.

FILL IN THE BLANKS

1. Profit is often determined by _____ _____, _____ _____, and _____ _____ of the animals.

2. Extremely heavy muscled pigs are associated with a condition known as _____ _____ _____.

3. Females that are too heavily muscled are less _____ and have problems _____.

4. The _____ _____ is a cross section of the long muscle that runs down both sides of the backbone of the pig.

5. _____ refers to the physical characteristics of the animal and _____ refers to the actual genetic makeup of the animal.

6. From a side view, a fat, light muscled pig will appear _____, particularly at the _____ _____ and at the base of the _____.

7. A properly muscled and conditioned pig should display _____ _____ in the hams.

8. The grade is an indication of the amount and quality of _____ _____ that can be expected from the _____.

9. Quality refers to the _____ of the pork and includes such measures as firmness of the _____ and thickness of the _____ _____.

10. A pig that has less than _____ _____ of belly fat is considered unacceptable. The belly area is where _____ comes from and a belly with fat coverage less than _____ is considered _____ _____.

MULTIPLE CHOICE

1. Pigs were first raised to produce mostly:
   a) hams
   b) chops
   c) lard
   d) pets

2. Selection efforts culminated in the 1960s with the:
   a) huge pig
   b) compact pig
   c) average pig
   d) super pig

3. Females with large, bulging muscles:
   a) are desirable
   b) have problems conceiving
   c) have problems delivering piglets
   d) both b and c

4. The loin eye is:
   a) the cross section of the muscle running down both sides of the backbone
   b) the top of the ham
   c) where bacon comes from
   d) should be ignored

5. In the 1990s, the trend was toward pigs that:
   a) were long and very tall
   b) were lean and muscular
   c) had large amounts of back fat
   d) had huge, bulging hams

6. A modern market hog should weigh about:
   a) 230 pounds
   b) 190 pounds
   c) 340 pounds
   d) 270 pounds

7. When viewed from the rear, a well muscled market pig should be:
   a) widest through the center of the hams
   b) widest at the top
   c) widest at the bottom
   d) round and bulging like a basketball

8. The back fat thickness on a US No. 1 pig should be:
   a) more than 1 inch
   b) less than 1 inch
   c) no more than 1.5 inches
   d) at least 2 inches

9. The expected yield of a US No. 2 pig is:
   a) between 57.4% and 60.3%
   b) over 70%
   c) less than 50%
   d) between 40.5% and 60%

10. When grading market hogs, the back fat measurement is taken:
    a) over the first rib
    b) over the fourth rib
    c) over the last rib
    d) over the middle rib

11. All market barrows and gilts which may produce soft or oily pork will be graded:
    a) US No. 4
    b) US Utility
    c) US No. 3
    d) US No. 5

 DISCUSSION

1. List the factors that often determine the profit made from a swine operation.
2. What brought about the trend away from raising pigs mostly for lard?
3. Describe a pig suffering from porcine stress syndrome (PSS).
4. Explain why extremely heavy muscled pigs have problems reproducing.
5. Why is the measurement of loin eye important?
6. List three ways to tell a fat pig from a lean pig.
7. Describe how a muscle bound pig walks.
8. What criteria are used to grade market pigs?
9. List the characteristics of a US No. 1 market pig.
10. Describe a US Utility market pig.

CHAPTER 6

Using Performance Data and Expected Progeny Differences in Swine

OBJECTIVES

As a result of studying this chapter, students should be able to:

- Recognize differences between performance data, indexes, and expected progeny difference data as used for selecting swine
- Describe key data terminology and the importance for using data when evaluating swine data
- Describe how performance data and expected progeny differences are used in judging contests and in production agriculture in relation to swine
- Compare and rank a set of animals given a specific scenario and data to match needs of the scenario
- Describe differences between terminal and maternal breeding characteristics in producing swine

PERFORMANCE DATA

Performance data is an important part of the swine industry and an integral education tool for teaching students about the importance of herd selection process using data. Producers use performance data to improve carcass, growth, and reproductive traits. Producers raise livestock to make money, so they focus on traits that will have the most impact on the herd depending

KEY TERMS (continued)

- Number born alive (NBA)
- 21-day litter weight (LWT)
- Number weaned (NW)
- Back fat (BF) ratio
- Pounds of lean
- Loin eye area (LEA)
- Scenario

on production needs and marketing opportunities for herd **progeny** (offspring). **Performance data** (production data) is actual data recorded and submitted to breed associations by producers. Breed associations use the actual data submitted from producers to compile expected progeny differences.

Expected progeny difference (EPD) is the estimate of the genetic value of an animal in passing genetic traits to its offspring. EPDs are a measure of expected differences in performance of a sire's (father's) or dam's (mother's) progeny by comparing individual animals to the average progeny of all sires and dams within the same breed (Figure 6-1). The prediction (EPD) is based on actual performance, progeny performance, and relatives' performance. Students should compare given data to the breed average while evaluating EPDs. The **breed average** is an average of all current EPDs for a specific breed. Breed averages change over time as new production data is added to the breed database. Breed EPD averages should only be compared to the breed average for the specific breed. EPDs should not be compared across breeds. EPDs are predictions of what is expected to happen with the offspring. This is based on a combination of action data from the individual animal, then entered into a database that combines that information and takes into consideration the relatives of the animal.

Production data is actual information from specific animals recorded by producers raising livestock. Performance data represents actual numbers from a producer's herd; these numbers are not estimates or predictions. Purebred producers submit production data on registered animals to the respective breed associations. Breed associations use the production data to calculate the breed averages and estimates to develop EPDs.

In order to give students hands-on opportunities to learn how to evaluate EPDs and performance data, livestock judging contests offer classes for students to use the data in helping place the class. Two basic types of classes are offered

FIGURE 6-1 Expected progeny difference is the estimated genetic value of an animal. (© Blaz Kure, 2010. Used under license from Shutterstock.com.)

where performance data is used: placing class and **keep/cull**. Placing classes consist of four gilts or boars that students place with the assistance of performance data or EPDs. Keep/cull classes usually consist of eight gilts or boars, and students must decide which four to keep and which four to cull (eliminate from the group) with the assistance of performance data. The combination of visual and performance data in livestock evaluation better helps prepare students for realistic selection decisions.

UNDERSTANDING PERFORMANCE DATA

Understanding data used in livestock judging classes is one of the most important steps. Swine classes generally consist of three types of data: actual production data (individual animal performance records), ratios and indexes, and genetic merit estimates (EPDs). In order to be successful in judging contests, students must understand what each statistic means and the pros and cons of each as it relates to the class and the specific scenario. Knowing how to interpret the data given is essential to ranking the class in the correct order.

Actual production data or individual performance data is the simplest and least valuable performance information to use in a contest. These are simply an individual animal's records with no consideration of environmental factors or comparison to any other groups. Individual records are presented as actual and as adjusted values. **Adjusted** numbers are common in dealing with livestock EPDs or data in many different situations. When considering information on livestock, an understanding of variables is needed. For example, a farmer in Georgia may have a group of pigs of the same age and very close weights as a farmer in Iowa. Each farmer can decide to feed the entire group the same type of feed at the same time every time and the same amounts. Some may think this would be a good experiment to test which feed or which group of pigs is the best, but several different factors could affect the results. These factors are often called **variables** in research or science. For example, the outside temperatures, the breeding or genetics, the access to water, the inside air flow of the barn, the type and/or size of the pen are all variables that could cause the group to grow faster or slower and not necessarily be a reflection of the pigs or the feed (Figure 6-2).

There are always many variables to consider when viewing livestock numbers and data, so they are adjusted to try to bring all the data to a relatively common value or average. Adjusted data should be given more importance and relevance because it takes into consideration the effect of some of the variables affecting different groups. Individual performance should only be used if the whole class was raised together (born same time, fed same feed, etc). If actual performance data is the only data presented, then focus on adjusted data such as days, back fat scan, and loin eye area at 250 pounds.

Ratios and indexes are calculations that take into consideration how each animal performed relative to a certain group. They are better than actual performance data on individual animals because they are relative to a particular group of animals. The ratio average is always 100 for the group, so ratios above 100 are considered to be superior to the group. Ratios of more than one trait are often combined to form indexes. The most common indexes used in contests are **terminal sire index (TSI), maternal line index (MLI),** and **sow productivity index (SPI).**

FIGURE 6–2 The Environmental conditions under which the peaks are raised can have an impact on growth rate. *(Photo by Bob Nichols, USDA Natural Resources Conservation Service.)*

Terminal traits and *maternal traits* are terms used to help describe the expected use of the progeny. Indexes combine several EPDs into a single value for each animal. TSI ranks individuals for use in terminal crossbreeding programs. TSI combines EPDs for back fat, days to 250 pounds, pounds of lean, and feed/pound of gain relative to their economic values.

Terminal traits refer to indexes and EPDs relating to the hogs in terms of market readiness or carcass merit; that is, terminal animals generally are raised strictly for meat or market purposes. Terminal traits focus on EPDs that will have the most influence on improving growth, leanness, and feed efficiency. Terminal in most cases refers to the intention of the producer for the animals to be harvested and not to be used as breeding stock (Figure 6-3). One of the most important concepts for students to understand is **days to 250**, which refers to the ideal market weight for the average market hog. Most packers are set up for the average market hog to weigh 250 pounds. Most packers will accept market hogs to range from 230 to 280 pounds. Hogs lighter and heavier than this ideal market range are more challenging for packers to process and give more variety to carcass measurements and sizes. However, if the majority of the hogs weigh a certain weight, then packers can estimate that the majority of the carcasses or meat they market will be within a certain range. For example, if all the live market hogs they process weigh between 230 and 280 pounds, then the packers can estimate that most of the loin eye areas will also measure within a certain range. Therefore, the packers will be able to market this product to consumers with more certainty of what they want and create more precise demand for a more standardized product. Most importantly, days to 250 refers to the number of days the pig will be on feed. This is a very important EPD to consider for almost any scenario, because

FIGURE 6–3 Carcass merit is an example of a terminal trait. *(© Picsfive, 2009. Used under license from Shutterstock.com.)*

the number of days on feed affects all producers in terms of feed and production costs. The faster the pigs grow the quicker the producer can sell the pigs and make the money back.

Maternal line index is used as a selection tool for producing replacement gilts. MLI combines EPDs for both terminal and maternal traits relative to their economic values in a crossbreeding program, placing twice as much emphasis on reproductive traits as postweaning traits. Sow productivity index ranks individual animals for reproductive traits. SPI combines the EPDs for NBA, number weaned, and LWT relative to their economic values when used in a crossbreeding program. Maternal refers to indexes and EPDs relating to mothers' ability and influence for the offspring; some scenarios will refer to **maternal traits** as reproductive traits. Maternal traits are measured on the sow, regarding farrowing and raising a litter. Traits most commonly used in determining maternal ability are number born alive, number weaned, and 21-day litter weight (Figure 6-4).

EPDs are the best performance data to compare potential genetic differences between animals. They are the best to use and most accurate in determining the effect of the data on the group evaluating, because EPDs have already been adjusted statistically. EPDs are in actual units. The days to 250 EPDs are in days, back fat EPDs are in inches, number born alive EPDs are in number of pigs, pounds of lean EPDs are in pounds, and 21-day litter weight EPDs are in pounds. Positive EPDs are desired for number born alive, pounds of lean, and 21-day litter weight. Negative EPDs are desired for back fat and days to 250 pounds. The swine data management system is known as the **Swine Testing and Genetic Evaluation System (STAGES)**. Managed through Purdue University, STAGES was started in 1985 with additional help from the USDA, the National Association of Swine Records, and the National Pork Producers Council.

Number born alive (NBA) is the number of live pigs farrowed in a litter. NBA is adjusted for parity (number of times the sow farrowed a litter) of the sow. Daughters of sires with positive (+) EPD for NBA will farrow larger litters than

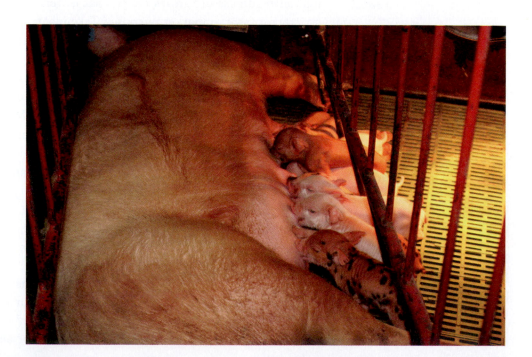

FIGURE 6-4 Traits most commonly used in determining maternal ability or number born alive, number weaned, and 21 day a litter weight. *(© Rick Whitacre, 2009. Used under license from iStockphoto.com.2009.)*

average EPD females. The **21-day litter weight (LWT)** is adjusted to 21 days of age. The LWT refers to the weight of the litter of pigs at the wean date. The industry standard and recommended days for weaning pigs is 21 days. Some producers wean earlier and some producers wean later than 21 days, but industry standard and most common is to wean 21 days from farrowing. Daughter of sires with positive (+) EPD for LWT will produce heavier litters than average EPD females. **Number weaned (NW)** is the number of pigs that a dam (mother) raised to 21 days of age (adjusted for parity and number after transfer). The EPD for NW is not reported, but it is used in the calculation of indexes.

As mentioned, days to 250 is the estimated days to reach 250 pounds. A National Swine Improvement Federation (NSIF) equation calculates days to 250 pounds from the animal's weight and age. Sires or dams with negative (-) EPDs will produce pigs that reach market weight faster than pigs of parents with average EPDs. **Back fat (BF) ratio** is a measure of back fat thickness. Actual back fat thickness is measured with an ultrasound in inches; the measurement is adjusted to 250 pounds live weight. Sires or sows with negative EPDs for back fat will produce pigs that have less back fat at market weight than pigs of parents with average EPDs. **Pounds of lean** is a measure of pounds of fat-free lean adjusted to a 185 pound carcass or approximately a 250 pound live pig. The pounds of lean is calculated from the back fat and loin eye area EPDs. A sire with a positive (+) EPD for pounds of lean will produce offspring that yield a higher percent of lean than offspring from a sire with a lower EPD for pounds of lean. **Loin eye area (LEA)** is measured in square inches, adjusted to 250 pounds live weight. The EPD for LEA is not reported but it is used in calculating pounds of lean.

EPDs provide predictions of how the progeny of a particular animal will compare to the progeny of other animals within a specific breed. EPDs are expressed as plus or minus values, with the average EPD for the population approximately zero (Figure 6-5). Remember for number born alive and 21-day litter weight, positive numbers are better; negative numbers are preferred for days to 250 and back fat.

FIGURE 6–5 EPDs provide a prediction of how the progeny of a particular animal compares to the progeny of other animals when a specific breed. *(Photo by Jeff Vanuga, USDA Natural Resources Conservation Service.)*

ANIMAL NUMBER	DAYS TO 250 EPD	BACKFAT EPD
1	−3.0	−0.05
2	0	0.05

FIGURE 6–6 Progeny of animal 1 is expected to reach 250 pounds three days sooner than progeny of animal 2.

ANIMAL NUMBER	EPD FOR 21-DAY LITTER WEIGHT	EPD FOR NUMBER BORN ALIVE
1	+2.0	+1.5
2	0	0

FIGURE 6–7 Progeny (daughters) from animal 1 is expected to produce 1.5 MORE pigs born alive per litter.

In figure 6-6, progeny of animal 1 is expected to reach 250 pounds three days sooner than progeny of animal 2. Progeny of animal 1 is also expected to have 0.1 inch less back fat (when bred to pigs with similar genetic background).

In figure 6-7, progeny (daughters) from animal 1 is expected to produce 1.5 more pigs born alive per litter. Progeny from animal 1 is also expected to have litter weights 2 pounds heavier at 21 days than animal 2 (when bred to pigs with similar genetic background). Accuracy (ACC) is an important part of understanding EPDs. Accuracy levels range from 0 to 1.0. The closer the ACC gets to 1.0 the better the EPDs are considered. The lower the ACC, or closer to 0, the more unreliable the EPD is considered. ACC levels are generally not given to students in contests; however, producers should consider ACC levels carefully when choosing sires or dams for their herd.

INTERPRETING AND EVALUATING SCENARIOS

Every class with data will have a **scenario** (or situation). The students should place the class based on the objectives of the scenario. The scenario is a brief overview of what the producer wants to happen from using the data and the producer's objectives for producing offspring. For example, some scenarios will ask students to judge the class looking for terminal traits or maternal traits. If the scenario suggests focusing on terminal traits, then emphasis should be placed on EPDs for growth, leanness, and muscling (days to 250, back fat, loin eye area, and/or percent lean). If the scenario focuses on maternal traits, then students should emphasize and place the class based on EPDs that may affect reproduction or mothering ability of offspring (number born alive, 21-day litter weight, SPI, and/or MLI). Before reading the scenario, students should note class description (breed) and briefly note whether EPDs, actual data, or ratios are given. The student must pretend to be the producer and rank the animals in order from the animal that best fits what the scenario suggests. For example, the scenario from the 2005 national FFA CDE swine keep/cull class is shown in table 6-1.

TABLE 6-1: 2005 National FFA CDE Swine Female Selection

KEEP / CULL HAMP X YORK X LANDRACE GILTS
PRODUCTION DATA SHEET

These crossbred gilts will be bred to Duroc Boars in a Terminal Crossing System. All Hogs will be raised in a total confinement unit with marketing taking place on a % lean valued system.

Within the scenario and data heading, students should be able to determine the following:

Class name: Keep/cull Hamp x York x Landrace Gilts

Class purpose: Describes the planned use of the animals—gilts will be bred to Duroc Boars

Production conditions: Describes the production environment, which is total confinement. Total confinement refers to hog operations where the hogs are grown completely inside a barn. The flooring is usually concrete or slatted floors (concrete or wire floors with slats). Total confinement indicates that pigs must be structurally correct or able to withstand being on those types of floors with relatively little room for walking or exercising.

Marketing intentions: Describes plan for selling animals for the specific class or their progeny—all hogs will be sold on a percent lean value system.

While reading scenarios, students should circle key words that help them know which traits to emphasize either visually or through performance data. For example, if the scenario states "all hogs will be raised in confinement," then soundness is important. Any hogs with major structure problems should be noted and possibly placed down in the class if the structure is severe or could affect production of the animal. Other examples are as follows:

- "All hogs will be sold on a pound of lean system" indicates muscling and leanness are important (visually) and that EPDs for pounds of lean and back fat are important.

- "This operation markets show pigs" indicates growth data is important and visually hogs should be balanced and have eye appeal for the show ring.

USING PERFORMANCE DATA IN ORAL REASONS

It is important to incorporate performance data into reasons for classes with data. Analysis of performance data should be added to the reasons in simple phrases that show the reason taker you understand the data and why you placed emphasis on certain traits. The reasons should combine visual descriptions of the class as well as why certain animals are considered to have more favorable EPDs in certain areas. Students should follow the direction of their adult leader for reasons format and terms. Students should be familiar with reasons format and terms before adding performance data. Once students understand reason terms and performance data, adding descriptions of performance data to the reasons is fairly simple. Some sample phrases to use in starting reasons or in top pair are, "I placed this class of Duroc Gilts 1-2-3-4. I started the class with the soundest structured, maternal design gilt who excels the class in maternal and growth genetics," or "Based on the scenario given which emphasized soundness, maternal excellence, and genetics, I placed the class of Duroc Gilts 1-2-3-4." The scenario and performance data should help students rank the class. It is very important to use the data and scenario in reasons from these classes. Other examples of how to incorporate performance data into reasons are as follows:

- "The boar that offers a lower back fat EPD and should stay leaner to a higher weight also has more lower skeletal width, and more muscle down his top and into his ham."
- "Bigger scale, wider based, more production oriented gilt who has an advantage in sow productivity index."
- "A sounder structured gilt who is more flexible out of both ends of skeleton should be more adjustable to the confinement scenario."
- "More maternally designed gilt who excels in maternal line index."

A simple way to add performance data to your reasons is to determine the actual difference between two animals (in a pair), subtract the lowest from the highest to determine an actual number, and add to reasons terms when discussing the pair. For example, consider the following:

- "Boar 1 is 0.5 inch leaner in his back fat EPD." (Subtract specific animal EPDs from other animal discussing in pair.)
- "Boar 2 has an advantage of 3.7 days in his days to 250 EPD."
- "Gilt 2 should produce progeny with litter weights 3.2 pounds heavier."

Be flexible with your reasons, understand the data, and add simple phrases into your reasons as much as possible. Other examples are as follows:

- "Gilt 1 excels the class with the lowest day to 250 EPD in the class."
- "Boar 2 is the lightest muscled, smallest framed boar, and has the slowest days to 250 pounds in the class."
- "I placed gilt 4 last in this class as she is the lightest muscled, narrowest made gilt with the poorest genetic evaluation in the class."

SUMMARY

Judging classes with performance data is one of the most educational parts of livestock judging contests. They prepare students for real life experiences in a fun, competitive environment. Performance data should help students to make decisions on classes. In order to be successful with performance data classes, follow these few steps:

1. Understand and be able to interpret production data, ratios, indexes, and EPDs for all swine breeds.
2. Understand and be able to analyze scenarios to find class name, class description, and producer objectives. Set priorities for the class based on scenario objectives: Know data is important and what physical traits are important before ever looking at animals.
3. Evaluate performance data, rank class on paper.
4. Evaluate visual traits.
5. Decide on final ranking, which should be a combination of visual and performance data.

Students should always read and evaluate the scenario before looking at the animals to place. As a student, make sure you understand the scenario objectives and data and have an initial paper placing in mind before visually evaluating the class. A rule of thumb for placing classes with performance data is that the data should help students pair the class. Usually two pigs are superior on paper and two pigs are inferior on paper, as long as you understand the data. You can usually get the pairs correct. It is okay to make a pair switch after you look at the animals. For example, based on the data, you may decide you definitely like the data for 2 and 3, but you do not like the data for 4 and 1. After you look at the class, however, you think 3 is better than 2, so your final placing might be 3-2-1-4. It all depends on the scenario. Contest officials are not trying to trick you. They want the contests to be a positive learning experience for every student.

STUDENT LEARNING ACTIVITIES

1. Visit a farm that uses data or have the teacher make up data to match four animals so students can evaluate the animals along the data and place the class.
2. Watch a video or view realistic pictures of animals and compare to data.
3. Order a catalog from different boar studs. Use the pictures and data to compare and choose boars to use on your own SAE breeding gilt project; or use the catalog to do a class presentation on how to choose boars from the catalog.
4. Use the catalog to develop your own class of boars using the data from different boars in the catalog in a worksheet format. Share with your class and teacher to see which boar the class would pick. (Most of this information can also be found online at different boar stud and breeder websites).

TRUE/FALSE

1. Progeny refers to the ancestors of the current generation.
2. Breed associations use the actual data submitted from producers to compile EPDs.
3. Actual production data or individual performance data is the most complex and the most valuable performance information to use in a contest.
4. If actual performance data is the only data presented, then students should focus on adjusted data more than actual unadjusted data.
5. TSI stands for terminal sire index.
6. EPD stands for expected progeny data.
7. LWT stands for light weight pigs.
8. Structural correctness is important for pigs raised in total confinement.
9. The pounds of lean is calculated from the back fat and loin eye area EPDs.
10. Days to 250 means that pigs will be sold at 250 days of age which is ideal market age.

FILL IN THE BLANKS

1. _____ is an important part of the swine industry and an integral education tool for teaching students about the importance of herd selection process.

2. Producers use performance data to improve _____, _____, and _____ _____.

3. EPDs (_____ _____ _____) estimate the genetic value of an animal in passing genetic traits to its offspring.

4. _____ _____ _____ refers to the ideal market weight for the average market hog.

5. _____ _____ _____ combines the EPDs for NBA, number weaned, and LWT relative to their economic values when used in a crossbreeding program.

6. _____ EPDs are desired for number born alive, pounds of lean, and 21-day litter weight.

7. _____ EPDs are desired for days to 250.

8. _____ _____ _____ is a measure of pounds of fat-free lean adjusted to a 185 pound carcass or approximately a 250 pound live pig.

9. The industry standard and recommended days for weaning pigs is _____ days.

10. _____ _____ in scenarios indicates to the students that pigs must be structurally correct.

DISCUSSION

1. Describe the importance of performance data in the livestock industry.

2. Describe the differences between EPDs, actual data, and ratios and indexes.

3. What is the difference between maternal and terminal traits?

4. Describe the difference between the SPI and MLI.

5. Why is days to 250 an important EPD for all producers?

6. Describe why number born alive would be important to a producer.

7. What physical traits should be examined closely for scenarios, including total confinement for the hog operation?

8. Why should producers use adjusted data rather than actual data when available?

9. In comparing four gilts in a scenario, why should you choose the gilt with the highest 21-day LWT over three other gilts with much lower 21-day LWT?

10. If a particular scenario states that the producer markets the pigs on a lean value system, which EPDs would be most important?

REFERENCES

Cleveland, Erik, and Todd See. *Swine Genetics Handbook, Fact Sheet 15: Selection Programs for Seedstock Producers.* West Lafayette, IN: National Swine Improvement Federation, Purdue University.

Cooperative Extension Service, *www.nsif.com/handbook.htm.*

Definitions of Terms Used by STAGES. Purdue University Animal Science Department, *www.ansc.purdue.edu/stages/glossary.htm.*

Formulas for STAGES Indexes. Purdue University Animal Science Department, *www.ansc.purdue.edu/stages/indexes.htm.*

Lipsey, Jerry, Marty Ropp, and Kyle Rozeboom. *Swine Genetics Handbook, Fact Sheet 17: Understanding and Using Performance Data in Judging Classes.* West Lafayette, IN: National Swine Improvement Federation, Purdue University Cooperative Extension Service, *www.nsif.com/handbook.htm.*

SECTION THREE

CATTLE

CHAPTER 7
Selecting Breeding Cattle

KEY TERMS

- Feed efficiency
- Frame size
- Reproductive (capacity) efficiency
- Sex character
- Structural soundness
- Post legged
- Sickle hocked
- Splayfooted
- Buck kneed
- Condition

OBJECTIVES

As a result of studying this chapter, students should be able to:

- Describe the trends in the beef industry
- Explain why the trends have changed
- Describe the modern type of beef animal
- Distinguish between animals that are structurally correct and those that are incorrect
- Explain the meaning of femininity and masculinity
- Select animals with the correct sex character
- Distinguish between animals that have the correct condition and those that are too fat or too thin

CORRECT TYPE OF BREEDING ANIMAL

The important end point of the entire beef industry is the retail cuts bought and consumed. As detailed in the next chapter, consumers demand a certain type of product to serve for meals. This means that the retail outlets must stock and sell the type of beef consumers want. In turn, the packers must supply what the retailers demand and the feedlot operators must deliver the type of animals the packers want for slaughter, figure 7-1. On up the chain, the producers must supply the feedlot operators with calves that will grow out and finish into the type of animal the feedlot operators want.

FIGURE 7–1 Feedlot producers must deliver the product wanted by the packers and the packers must deliver the products wanted by the consumer. *(Courtesy of Natural Resources Conservation Service. Photo by Jeff Vanuga.)*

The beginning point is the production of the type of breeding animals that will reproduce and supply the proper type of calves. Over the years, what was considered the correct type of animal has changed considerably. In the 1950s, the preferred type of animal was very short legged and blocky in appearance. They were often described as looking like a bale of hay with short legs. In fact, photographs of winning show animals often showed animals standing in hay to make them appear shorter. The problem with this type was that they finished at an early weight and often had too much fat that was scrapped as wastage. In the 1960s, selection began for animals that were large framed and that finished at a much larger size. The thinking at the time was that these animals grew faster on less feed per pound of gain. What became known as the exotic breeds were brought in to meet these selection criteria. Breeds such as Charolais, Simmental, Limousin, and Main Anjou were used to produce longer legged, larger framed calves. By the end of the 1960s, steers were produced that finished at 1,500 to 1,600 pounds.

This size resulted in two problems. First, consumers were reluctant to buy steaks that weighed more than a pound. Rib eyes could reach as much as 20 square inches and this proved too much for the average consumer to eat at a serving. The second problem was in **feed efficiency**. Research showed that there was really little difference in feed efficiency (weight gained per pound of feed consumed) if the animals were fed to the same level of fatness. A small steer that graded low choice had as efficient weight gain as a large frame steer that was fed to a low choice grade. These problems led to the selection of animals that would finish at around 1,000 to 1,200 pounds, figure 7-2. Obviously, breeding animals are selected for this type of animal.

FIGURE 7–2 The desirable steer finishes at about 1,000 to 1,200 pounds and grades a low choice. *(Courtesy of Dr. Frank Flanders.)*

Not only must producers produce the correct type of calf, but they must also be able to make a profit. This means that the animals in their herd must be able to reproduce efficiently and raise calves that will grow rapidly and wean at a sufficient size. In evaluating breeding animals, several factors must be considered. These factors include frame size, reproductive capacity, structural soundness, and production records.

FRAME SIZE

Frame size refers to the overall height of the animal at maturity; tall animals are larger framed than short animals. This measurement is important because the consumer wants a medium size cut of beef instead of a very small or very large cut. This means that packers of beef prefer carcasses that weigh between 600 and 700 pounds. Frame size is determined by measuring the animal at the hip at a certain age. Figure 7-3 shows the frame size as indicated by the hip height when corrected by age.

Frame score can also be estimated in a young calf by measuring the cannon bone, which lies between the knee and the ankle on the front legs. Research has shown that an animal with a longer length of cannon bone will be a taller animal at maturity than an animal of the same age with a shorter cannon bone. In fact, some breed associations request that the producer record the length of cannon at birth of any animals they wish to register. Animals with a frame score of approximately 5 (medium frame) are preferred because calves of this size should finish at around 1,000 to 1,200 pounds.

REPRODUCTIVE CAPACITY

Reproductive capacity or **efficiency** means that cows must be able to produce and raise a healthy calf at least once per year. If the cow does not become pregnant shortly after her calf is raised, time and effort is lost. This can be caused by a cow that does not conceive because of reproductive problems or a bull that is incapable of efficiently impregnating the female. Several indicators of reproductive efficiency can be observed in live animals.

FRAME SCORES
MALES

Frame Scores Based on Height (in inches) Measured at Hips

Age in Months	Frame Score 2	Frame Score 3	Frame Score 4	Frame Score 5	Frame Score 6	Frame Score 7
5	36.00	38.00	40.00	42.00	44.00	46.00
6	37.00	39.00	41.00	43.00	45.00	47.00
7	38.00	40.00	42.00	44.00	46.00	48.00
8	39.00	41.00	43.00	45.00	47.00	49.00
9	40.00	42.00	44.00	46.00	48.00	50.00
10	41.00	43.00	45.00	47.00	49.00	51.00
11	42.00	44.00	46.00	48.00	50.00	52.00
12	43.00	45.00	47.00	49.00	51.00	53.00
13	43.50	45.50	47.50	49.50	51.50	53.50
14	44.00	46.00	48.00	50.00	52.00	54.00
15	44.50	46.50	48.50	50.50	52.50	54.50
16	45.00	47.00	49.00	51.00	53.00	55.00
17	45.50	47.50	49.50	51.50	53.50	55.50
18	46.00	48.00	50.00	52.00	54.00	56.00
19	46.25	48.25	50.25	52.25	54.25	56.25
20	46.50	48.50	50.50	52.50	54.50	56.50
21	46.75	48.75	50.75	52.75	54.75	56.75
22	47.00	49.00	51.00	53.00	55.00	57.00
23	47.25	49.25	51.25	53.25	55.25	57.25
24	47.50	49.50	51.50	53.50	55.50	57.50

FEMALES

Frame Scores Based on Height (in inches) Measured at Hips

Age in Months	Frame Score 2	Frame Score 3	Frame Score 4	Frame Score 5	Frame Score 6	Frame Score 7
5	35.75	37.75	39.75	41.75	43.75	45.75
6	36.50	38.50	40.50	42.50	44.50	46.50
7	37.25	39.25	41.25	43.25	45.25	47.25
8	38.00	40.00	42.00	44.00	46.00	48.00
9	38.75	40.75	42.75	44.75	46.75	48.75
10	39.50	41.50	43.50	45.50	47.50	49.50
11	40.25	42.25	44.25	46.25	48.25	50.25
12	41.00	43.00	45.00	47.00	49.00	51.00
13	41.75	43.75	45.75	47.75	49.75	51.75
14	42.25	44.25	46.25	48.25	50.25	52.25
15	42.75	44.75	46.75	48.75	50.75	52.75
16	43.25	45.25	47.25	49.25	51.25	53.25
17	43.75	45.75	47.75	49.75	51.75	53.75
18	44.25	46.25	48.25	50.25	52.25	54.25
19	44.50	46.50	48.50	50.50	52.50	54.50
20	44.75	46.75	48.75	50.75	52.75	54.75
21	45.00	47.00	49.00	51.00	53.00	55.00
22	45.00	47.00	49.00	51.00	53.00	55.00
23	45.25	47.25	49.25	51.25	53.25	55.25
24	45.25	47.25	49.25	51.25	53.25	55.25

The height under inches shown under each frame size is the minimum height for that frame size

FIGURE 7–3 Frame size is determined by the height of the animal's hip at a certain age.

In heifers, consideration should be given to those animals having a longer length between the hooks and pins and those that are wider apart at both hooks and pins. Distances should be wide from hook to hook, pin to pin, and long from hooks to pins, figure 7-4. This is an indication of a greater pelvic capacity. When

(A)

(B)

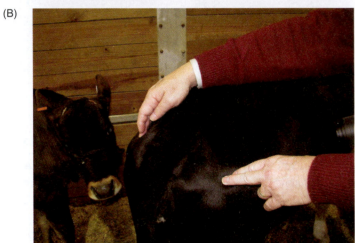

(C)

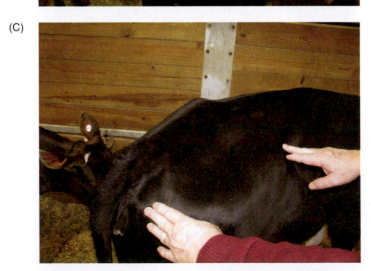

FIGURE 7-4 Total pelvic capacity is determined by the distance between the hooks (A), the pins (B), and from hooks to pins (C). *(Courtesy of Dr. Frank Flanders.)*

FIGURE 7–5 This heifer has a head like a steer and is thus termed a steer headed heifer. *(Courtesy of Shutterstock.)*

FIGURE 7–6 This heifer has adequate body depth. Notice the length and spring of the ribs. *(Courtesy of American Charolais Association.)*

a female gives birth, the pelvis must open enough to allow the passage of the calf through the birth canal. If a heifer has a small pelvis, she will probably have problems delivering a calf. The fertile female should be well balanced and present a graceful feminine appearance. She should be long and clean in the face and throat. Her neck should be long and blend smoothly into smooth, sharp shoulders.

A heifer that has a head shaped like a steer may have fertility problems, figure 7-5. She should be clean and trim through the brisket and middle. The pelvic area should be large and wide for easy calving. Breeding cattle should also have adequate body depth and width to provide adequate room for the internal organs. The larger the internal organs, such as the heart and lungs, the better the animal should do in terms of viability and growth. Indicators of capacity are width through the chest floor, long and well-arched ribs, and depth in the side, figure 7-6.

Sex character simply means that a bull looks like a male and a heifer looks like a female. Because sex hormones control both the physical appearance of animals and their ability to reproduce, it stands to reason that an animal with more sex character should be more reproductively efficient. Sex character in a bull is determined by a broad, massive, bull-like head. The shoulders should be bold and well muscled, but care should be taken that bulls with coarse, excessively thick shoulders are not selected as a herd bull. The shoulder blades should be set smoothly into the shoulder and not protrude like the bull in figure 7-7. Because this characteristic is passed on to the offspring, calving difficulties can be encountered.

One of the most important physical traits of a bull is that of testicle shape and size. A two-year-old bull should have a scrotal circumference of at least 34 centimeters when measured at the largest part. Research has shown that the larger the testicles, the larger the number of valuable sperm produced. In addition, the scrotum should not be tight and short. It should let the testicles extend

FIGURE 7–7 The shoulder blades on this bull protrude from the shoulder too much. Also note that the muscling in the hind quarter is long, thick, and smooth. This type muscling is desirable. *(Courtesy of Shutterstock.)*

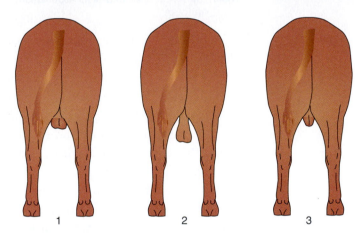

FIGURE 7–8 The testicles extend downward to about hock level and should have a definite neck to the shape. Bull 2 is the best choice. *(Courtesy of Vocational Materials Services, Texas A & M University.)*

downward to about hock level and should have a definite neck to the shape, figure 7-8. If the testicles are held too close to the body, then the temperature will be too high for ideal sperm production. The bull in figure 7-9 has a straight-sided scrotum that is too small for his age and the testicles are held too close to the body. The bull in figure 7-10 has the proper size and shape testicles.

STRUCTURAL SOUNDNESS

Structural soundness refers to the correctness of the feet and legs of an animal. The legs should fit squarely on all four corners of the animal. The correct set to the back legs of a beef animal is shown in figure 7-11. In a structurally correct animal,

FIGURE 7-9 This bull has testicles that are too small and carried too high. Also note that his head is small for his size. This bull was semen tested and proven to be not very fertile. *(Courtesy of Dan Rollins.)*

FIGURE 7-10 The testicles on this bull have adequate size and shape. *(Courtesy of American Gelbvieh Association.)*

SET OF LEGS AND FEET

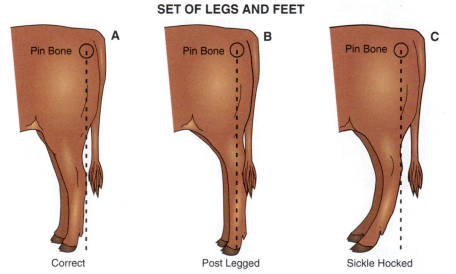

FIGURE 7-11 This diagram depicts animals that are correct, post legged, and sickle hocked. *(Courtesy of Vocational Materials Service, Texas A & M University.)*

FIGURE 7–12 This heifer is post legged.

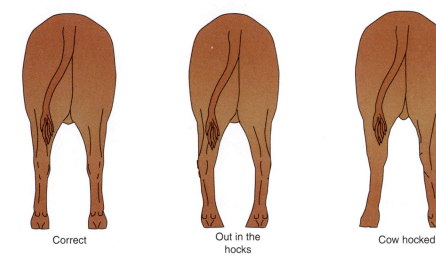

COMPARISON OF CORRECT WITH DEFECTIVE HIND LEGS

Correct

Out in the hocks

Cow hocked

FIGURE 7-13 This diagram shows animals that are correct, out in the hocks, and cow hocked. *(Courtesy of Vocational Materials Service, Texas A & M University.)*

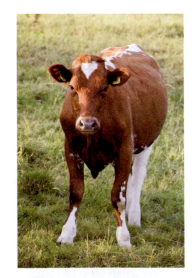

FIGURE 7–14 This heifer is splayfooted and buck kneed. *(Image copyright Sofia Kozlova, 2009. Used under license from iStockphoto.com.)*

a plumb line dropped from the pins to the ground will intersect with the hock. The heifer in figure 7-12 stands too straight on her back legs. This condition is known as **post legged**. The rear legs are too straight and do not provide enough cushion and flex as the animal walks. In bulls, this condition causes problems in mounting cows. The opposite condition, known as **sickle hocked**, also causes problems in mating. As the animal mounts, undue stress is placed on the stifle muscle, causing the animal to become stifled; that is, the ligament attaching the stifle muscle tears. This results in the animal being worthless as a herd bull. When viewed from the rear, the back legs should be straight. Figure 7-13 depicts animals that have structural problems as viewed from the rear. Figure 7-14 depicts animals that are structurally incorrect on their front legs. All cattle are slightly splayed in the front, but the front feet should turn out very little. If the front feet point in opposite directions, a condition known as **splayfooted** exists. This condition is not desirable because excess stress is placed on the knees and joint wear can lead to discomfort to the animal. When viewed from the side the knees should be straight. If they are bowed out the condition is called **buck kneed**. This condition also puts undue stress on the knee joints.

All animals should move out with a free, easy stride. The rear foot should be placed about where the front foot was picked up. Cattle that take short, choppy steps are either too tightly wound in their muscle pattern or have problems in their skeletal makeup. Either condition is objectionable. To feel their best and to grow and do their best, animals must be structurally correct.

CONDITION AND MUSCLING

Breeding animals must also have the correct amount of condition. **Condition** refers to the amount of fat a breeding animal is carrying. If an animal is too thin, fertility may be a problem. In order for the reproductive process to take place, animals must have sufficient storage of energy to not only maintain life but also allow the reproductive systems to function properly. If an animal is too fat, the act of breeding may be difficult. Also, females that have too much fat surrounding the internal reproductive organs may have problems functioning. Excessive deposits of fat can also restrict the birth canal during calving. Also keep in mind that a cow must produce enough milk to keep the calf healthy and growing rapidly. The number one factor in weaning heavier calves is the amount of milk the mother gives. A cow that is overly fat may not produce sufficient milk. The animal's body may be more efficient at producing fat than producing milk.

When observing a live animal, look for fat deposits in the flank area, the brisket, and the ribs. An animal that is too thin will have ribs that are visible from the backbone to the end of the ribs. A correctly conditioned animal will have some fat deposits at the top of the ribs, but some of the ribs may be visible as the animal walks. A heifer or bull that has a large, extended brisket is probably too fat. There should be enough fat to cover the bone, but not an excessive amount. The flank should be trim and not full and extended. Scientists have worked out a way of scoring the amount of fat on the body of a beef animal. This is known as the body conditioning score (BCS), figure 7-15.

Breeding animals should have sufficient muscling. If the sire and dam do not have adequate muscling, it stands to reason that the offspring will not have the correct amount of muscling. Both bulls and heifers should stand wide in both the front and rear. Narrow standing animals do not have adequate muscling. The muscles should be long and smooth, especially on heifers, figure 7-16. Round, bunchy muscling can cause problems by restricting the birth canal. Notice how the animal walks. As mentioned, the rear foot should be placed on the ground where the front foot was picked up. Tightly wound animals (too much muscling) will take short, choppy steps and will exhibit goose stepping. Tightly wound bulls will have problems moving around and will have trouble mounting in the mating process, figure 7-17. Also, calves from this type of bull may be too large to easily pass through the birth canal.

Thin

Moderate

Fat

BCS 4: Borderline condition. Outline of spine slightly visible. Outline of three to five ribs visible. Some fat over ribs and hips.

BCS 5: Moderate, good overall appearance. Outline of spine no longer visible. Outline of one to two ribs visible. Fat over hips but still visible.

BCS 9: Extremely fat, wasty, and patchy. Mobility possibly impaired. Bone structure not visible. Extreme fat deposits over ribs, around tail head and brisket. *(© 2009 iStockphoto.com)*

BCS 6: High moderate condition. Ribs and spine no longer visible. Pressure applied to feel bone structure. Some fat in brisket and flanks.

BCS 7: Good, fleshy appearance. Hips slightly visible but ribs and spine not visible. Fat in brisket and flanks with slight udder and tail head fat.

FIGURE 7–15 The body conditioning score (BCS) indicates the amount of fat on an animal's body. *(Courtesy of Dr. David A. Mangione, The Ohio State University.)*

FIGURE 7–16 This heifer has the correct type of long, smooth muscling. *(Delmar/Cengage Learning. Photo by Ray Herren.)*

FIGURE 7–17 The muscling on this animal is too round and bunchy. *(Image copyright Leslie Banks, 2009. Used under license from iStockphoto.com.)*

SUMMARY

Breeding animals should reflect the type of market animals needed. Medium framed animals that are structurally correct should be selected. Both bulls and heifers should have the proper condition and adequate muscling. Of paramount importance is the ability of breeding animals to reproduce efficiently. Animals that show the proper sex characteristics are more desirable because they are more likely to have the proper balance of reproductive hormones. Proper muscle structure, correct bone makeup, and the right amount of conditioning can directly affect birthing efficiency.

STUDENT LEARNING ACTIVITIES

1. Visit a producer and discuss what selection criteria is used when selecting breeding animals.
2. Visit the website of different breeds of beef cattle. List the conditions that can prevent an animal from being registered.

FILL IN THE BLANKS

1. Breeds such as Charolais, Simmental, Limousin, and Main Anjou are referred to as _____ _____.

2. In evaluating breeding animals several factors must be considered. These factors include _____ _____, _____ _____, structural soundness, and production records.

3. Frame score can also be estimated in a young calf by measuring the _____ _____.

4. Animals with a frame score of around _____ are preferred because calves of this size should finish at around _____ to _____ pounds.

5. If a heifer has a small _____ she will probably have problems delivering a calf.

6. _____ _____ simply means that a bull looks like a male and a heifer looks like a female.

7. If the _____ are held too close to the body, the temperature will be too high for ideal _____ _____.

8. _____ _____ refers to the correctness of the feet and legs of an animal.

9. _____ refers to the amount of fat a breeding animal is carrying.

10. Excessive deposits of fat can also restrict the _____ _____ during calving.

MULTIPLE CHOICE

1. Large framed animals are a problem because:
   a) they are too tall
   b) retail cuts from the animals are too large
   c) there is too much fat in the carcasses
   d) they grow too fast

2. When fed to a low choice grade, the feed efficiency for large frame animals is:
   a) greater than large frame animals
   b) less than small frame animals
   c) better than medium frame animals
   d) no different than any other frame size

3. Packers want market animals that will produce a carcass that will weigh:
   a) 1,000 to 1,200 pounds
   b) 500 to 600 pounds
   c) 800 to 900 pounds
   d) anything above 700 pounds

4. A two-year-old bull should have a scrotal circumference of at least:
   a) 34 cm
   b) 25 cm
   c) 100 cm
   d) 14 cm

5. If an animal's rear legs are too straight it is said to be:
   a) sickle hocked
   b) cow hocked
   c) post legged
   d) splayfooted

6. Splayed feet put stress on the:
   a) knees
   b) hips
   c) shoulders
   d) ankles

7. Short, choppy steps indicate:
   a) too much muscle
   b) structural problems
   c) either a or b
   d) no problems

8. A cow that is overly fat may:
   a) be just right
   b) have difficulty giving birth
   c) reproduce efficiently
   d) cost less to feed

9. Bulls and heifers should stand:
   a) wide in both front and rear
   b) narrow front and back
   c) narrow on the front and wide at the back
   d) narrow both front and rear

10. Muscling in heifers should be:
    a) as thick as possible
    b) round and bulging
    c) somewhat thin
    d) long and smooth

DISCUSSION

1. Trace the trends of the beef industry from the 1950s to the present.
2. Describe the proper size beef animal.
3. What is meant by reproductive efficiency?
4. What is meant by pelvic capacity?
5. Why is pelvic capacity important?
6. Explain why a heifer should look feminine.
7. What should a masculine bull look like?
8. Why is it important that a bull have structurally correct rear legs?
9. Analyze the problems encountered when a cow is too fat.
10. Explain the problems with a heifer that has too much muscle.

CHAPTER 8

Evaluating and Grading Market Beef Animals

OBJECTIVES

As a result of studying this chapter, students should be able to:

- Describe how the type of market beef animal has changed
- Explain consumer preferences in the cuts of beef they buy
- Explain why frame size is important in the selection of market animals
- Determine the proper amount of muscling on a beef animal
- Determine the proper amount of finish on a market beef animal
- Explain the difference between quality and yield grade
- Describe the process for quality grading a live animal
- Analyze the factors that go into quality grading a carcass
- Calculate the yield grade of a beef carcass

CONSUMER PREFERENCES

Americans have always been consumers of beef. Each year Americans consume about 65 pounds of beef per capita, figure 8-1. From the earliest part of our history, beef animals have been raised to provide meat for families to eat. Raising beef animals was generally considered to be done in small

KEY TERMS (continued)

- Commercial
- Utility
- Cutter
- Chronological age
- Physiological age
- Cartilage
- Ossification
- Pone fat
- Kidney, pelvic, and heart (KPH)
- Retail cuts
- Round

FIGURE 8–1 Each year Americans consume around 65 pounds of beef per capita. *(Image copyright Crossroads Creative, 2009. Used under license from iStockphoto.com.)*

operations up until the end of the American Civil War. Prior to that time most of the population of the country lived in rural areas where beef could be raised and slaughtered. By the middle of the 1800s, several large centers of populations had developed in the country. Cities such as New York, Boston, Philadelphia, and Chicago had relatively large numbers of people who had to be fed. At the end of the war, the cattle industry began to develop in the western states such as Texas and Wyoming. Large cattle drives were organized to move cattle from the range where they were grown to the population centers where they were slaughtered and consumed. Producers needed a very hearty, tough animal to survive on the range and to endure the long drives. Range cattle, such as the Texas Longhorns, were developed to withstand these rigors.

Cattle selection began to change with the invention of the refrigerated rail-car. The toughness in such breeds as the Texas Longhorn was no longer needed because the long drives to a market were no longer necessary. The cattle could be slaughtered near a railroad and the carcasses loaded onto the refrigerated rail-cars. Within a few days the beef could be sold in the large population centers. This development allowed producers to concentrate on producing faster growing, more efficient animals that could provide consumers with more desirable beef.

Since that time the type of beef animal selected for has changed. These changes are discussed in Chapter 7. One factor that has remained fairly constant in selecting market beef animals is consumer preference. Consumers play an important role in determining what type of beef animal is raised for slaughter. They usually want beef that is tender, flavorful, and affordable, figure 8-2. To produce this type of product, the right type of animal must be produced.

Tenderness and taste are both related to the age of the animal. Most of the beef sold in the supermarket as retail cuts comes from a choice grade. This means

FIGURE 8–2 Consumers want high-quality beef at a reasonable price. *(© Michael Krinke, 2009. Used under license from iStockphoto.com.)*

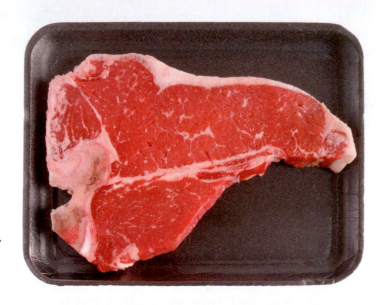

FIGURE 8–3 The small flecks of fat in this steak are the marbling that provides the flavor. *(© Marie-France Bélanger, 2009. Used under license from iStockphoto.com.)*

that the animal has reached a degree of maturity where it begins to deposit fat in the muscle. As animals grow and mature, fat is deposited differently. In young animals most of the energy from feed is put into the growth of bone and muscle so the animal can grow larger. When the animal matures, growth of bones and muscles ceases, and the animal begins to utilize energy from feed to deposit fat. Fat is first deposited in the body cavity of the animal. This serves not only to provide energy storage but also to help cushion the internal organs. As the body cavity reaches its peak in terms of fat deposit, the animal begins to deposit fat under the skin. When a certain level of back fat is reached, fat is deposited between the muscles and finally inside the muscles. These intermuscular fat deposits, called marbling, are what gives the meat its flavor, figure 8-3. Totally lean meat that is devoid of fat is dry and rather tasteless.

As will be discussed later in this chapter, carcasses are graded to a large degree on the amount of fat that is deposited in the muscles. Cattle, if fed correctly, usually reach the proper stage of fatness or finish when they are about two years of age. At this age, the animals are generally still young enough to be tender.

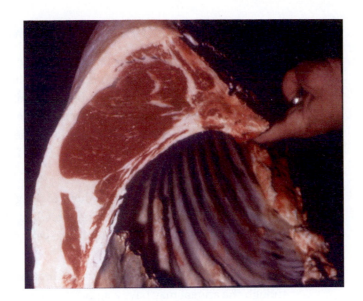

FIGURE 8-4 The largest muscle in a beef animal's body is the *longissimus dorsi* muscle that runs along each side of the backbone. The cross section is called the rib eye. *(Courtesy of Alvin Alford, Cooperative Extension Service, The University of Georgia.)*

The consumer does not want cuts of meat that have a lot of excess fat, so the idea is to produce animals that will put marbling into the meat with a minimum amount of back fat on the carcass.

Consumers are also choosy about the size of the meat cuts. If the animal is too large when slaughtered, the carcass may yield cuts that are too large. Consumers seem to want steaks that one person can consume in one meal. This means that carcasses that have a rib eye larger than 15 square inches may be too large for the average consumer. When the animal is slaughtered, the carcass is divided down the middle of the backbone. The long muscle running next to each side of the backbone is the **longissimus dorsi muscle**. This is the largest muscle in the animal's body and is an indication of the overall amount of muscle in the carcass, figure 8-4. Each half of the carcass is split between the twelfth and thirteenth ribs. The area of the cross section of the *longissimus dorsi* muscle is called the **rib eye**.

In order to be of the proper size at slaughter, animals have to be the right size when they begin to mature and lay down fat within the muscles. The size of the animal at maturity depends on the frame size of the animal. **Frame size** refers to the skeletal size of an animal at a given age. Small framed animals mature earlier than large framed animals, and they deposit fat in the muscle at a smaller size. A small framed animal (frame score 1) will probably grade choice between 750 and 850 pounds, while a large framed animal (frame score 7) will usually have to weigh 1,350 pounds or more to grade choice. The numerical score for frame size is determined by measuring the height of an animal at the hip at a certain age (see figure 7-3 in Chapter 7). Commercial packers usually want a carcass that weighs between 600 and 700 pounds. Carcasses within this weight range are more easily managed in the cutting room, and they provide the size of cuts that the consumer wants. Frame size should be large enough for the animal to grade choice at about 1,050 to 1,200 pounds in order to obtain the desired carcass weight and grade. This means that in selecting slaughter steers, preference should be given to the medium framed steers that finish at 1,050 to 1,200 pounds.

DETERMINING THE AMOUNT OF MUSCLING

The purpose in producing market beef animals is to obtain muscles that can be cut into **retail cuts** of beef for the consumer. It would seem that the more muscle there is on the animal, the more desirable that animal becomes. This is true only up to a point. Just as an animal can have too little muscling, it can also have too much. Selection for extreme muscling leads to the development of cattle with a condition known as **double muscling** (figure 8-5). Double muscling is undesirable for the following reasons:

1. These animals are difficult to produce. If the goal of the producer is to select animals with double muscling, then breeding stock of the same type must be selected. Fertility in these animals is very poor and calving is much more difficult.

2. The meat tends to be coarse and void of **intermuscular fat (marbling)**. Even though the animal may have a sufficient cover of fat, the marbling tends to be less than adequate. Thus, a large percentage of these animals grade standard.

3. Double muscled calves are difficult to raise because they are more susceptible to disease.

4. Double muscled feeder animals must be fed a higher proportion of concentrate in the ration in order to obtain enough marbling to grade choice. **Concentrate** feeds contain a high percentage of protein and carbohydrates which are generally supplied with grains such as corn. These cattle can be recognized by their physical appearance. The rump is protruding and round with definite grooves or creases between the thigh muscles. The tail is short and attached high and far forward on the rump. The head of the animal is small and long and is carried lower than the top of the body. Animals that possess an adequate degree of long, smooth muscling should be selected over animals that are lightly muscled or have extremely tight-wound, excessive muscling. **Smooth muscling**—that is, muscling that does not bulge too much—allows the animal to move freely and smoothly.

To judge a class of market beef animals, begin by comparing the animals for the degree of muscling. The place to start the comparison is to look at the animals

FIGURE 8–5 A double muscled animal is undesirable. The rump is protruding and round with definite grooves or creases between the thigh muscles. The tail is short and attached high and far forward on the rump. *(Courtesy of Getty Images/David.)*

through the center part of the round. This is the best indicator of overall muscling. The hind quarter should be the thickest through the center of the round, figure 8-6. An animal that is widest at the top is either too fat (overly finished), light muscled, or both. Also the animal should stand wide. Narrow standing animals are usually lighter muscled and should be placed lower in the class, figure 8-7. Animals that have small diameter bones are usually light muscled as well, so look for animals that have larger bone diameter. Keep in mind that the muscles are suspended from bones and light boned animals usually have light muscling. Look at the **twist** area where the animal's rear legs join. There should be a distinct cut-up appearance with a crease where the quarters join. If the area appears full and smooth with no clear cut between the quarters, the animal is either light muscled, overly fat, or both. The muscles in the hind quarter should be distinct and seams between the muscles visible. As the animal walks, the muscles should flex and be visible. A wide animal that shows no flexing of the muscles when it walks is probably too fat and should be moved down in rank.

Look at the animals from the side. The muscling should be carried throughout the body of the animal, figure 8-8. Be careful that you are observing muscling and not shoulder blades that stand out. Muscles should have ridges and there should be separation of the muscle groups. The top should be level and straight and have an indication of muscle particularly at the **loin area**. This is the part of the back from the hooks to the last rib. Remember that the most expensive cuts from the carcass come from this area and this part should be wide and deep. The side should be deep with ribs that appear sprung out. A flat-sided animal that has short ribs indicates an animal that is slow growing and lacking in body capacity.

Look down the top line of the animal. As in the rear view, the thickest part of the animal should be through the center of the hindquarter. An animal that appears to have an oval shape when viewed over the top is probably overly

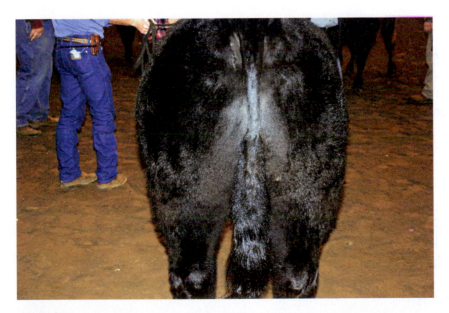

FIGURE 8–6 The muscling on the hind quarter of a steer should be the thickest through the middle as in this steer. *(Delmar/Cengage Learning. Photo by Ray Herren.)*

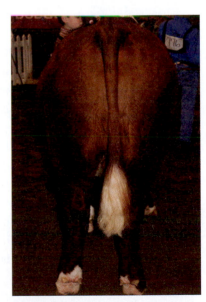

FIGURE 8–7 This animal is narrow standing and is thinly muscled. Note that the hind quarter is thickest at the top. *(Delmar/Cengage Learning. Photo by Ray Herren.)*

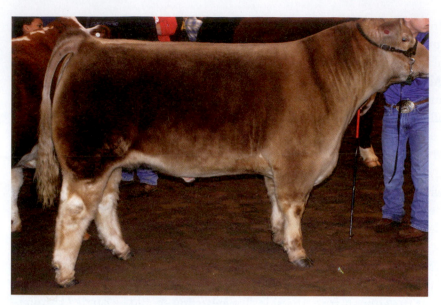

FIGURE 8–8 This steer is nicely balanced with the muscling carried throughout the body. Also the animal is carrying the proper amount of finish. *(Delmar/Cengage Learning. Photo by Ray Herren.)*

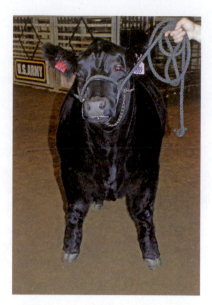

FIGURE 8–9 Steers should be muscled through the shoulders and stand wide like this animal. *(Delmar/Cengage Learning. Photo by Ray Herren.)*

finished and light muscled. There should be a visible distinction at the loin and the shoulder. There should be muscling over the shoulder that is indicated by seams of muscles that can be seen under the hide, particularly as the animal walks. Also observe the animal from the front. It should be muscled over the shoulder and stand wide, figure 8-9.

FINISH

When referring to a market beef animal, **finish** means the amount of fat on the animal. As mentioned, most market steers should have enough fat to grade choice. Choice is the grade most preferred by consumers. Once an animal matures and puts marbling in the muscle, the cost of putting more weight on the animal begins to increase dramatically. A properly finished animal will appear smooth over the rib cage, figure 8-10. Note that there are few muscles between the ribs and the hide. If you can feel the rib cage, there should be an even covering of fat over the ribs that should continue from the backbone to the end of the ribs. Practice will help you determine when there is too much of a cushion of fat over the ribs. Most times in a judging contest, you will not be able to feel the ribs and must make your judgment visually. Watch the animal as it walks. The ribs should not be visible and the cushion of fat should be apparent. Look at the brisket area. It should appear to be filled out but not bulging, figure 8-11. A bulging brisket indicates an animal that is overly finished. In the next section, the visual grading of beef animals will be discussed.

STRUCTURE

Market animals should also be structurally correct. Even though they are grown for slaughter, structurally correct market steers and heifers are more desirable and should be given an advantage in a judging contest. Animals that are structurally

TABLE 8-3 Shows the Degree of Marbling in each Category.

Relationship Between Marbling, Maturity, and Carcass Quality Grade[1]					
Degrees of Marbling	Maturity[2]				
	A[3]	B	C	D	E
Slightly Abundant	PRIME				
Moderate			COMMERCIAL		
Modest	CHOICE				
Small					
Slight	SELECT		UTILITY		
Traces					
Practically Devoid	STANDARD		CUTTER		

[1]Assumes that firmness of lean is comparably developed with the degrees of marbling and that the carcass is not a "dark cutter."
[2]Maturity increases from left to right (A through E).
[3]The A maturity portion of the Figure is the only portion applicable to bullock carcasses.

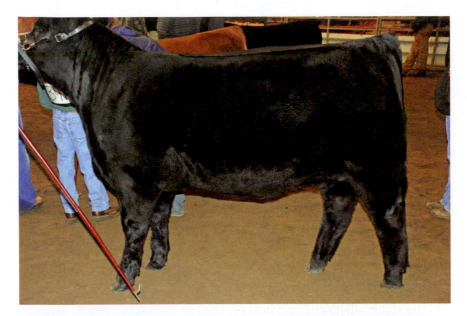

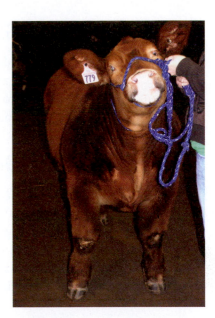

FIGURE 8–10 A properly finished market animal will appear smooth over the ribs. It should also feel spongy when handled. *(Delmar/Cengage Learning. Photo by Ray Herren.)*

FIGURE 8–11 The brisket on a properly finished steer should appear full but not bulging. *(Delmar/Cengage Learning. Photo by Ray Herren.)*

FIGURE 8–12 This steer is not structurally correct. He stands too straight on his hind legs and is post legged. *(Delmar/Cengage Learning. Photo by Ray Herren.)*

much fat has been deposited under the skin and between the muscles. A further complication is that different breeds may deposit intermuscular fat with varying amounts of subcutaneous (under the skin) fat. For example, some breeds may not start depositing marbling until they have 0.6 inch of back fat while others may begin depositing marbling at 0.35 inch of back fat.

The following guidelines are used by graders to determine quality grade in live animals. The guidelines are from the USDA publication, *Training Manual for USDA Standards for Grading Slaughter Animals*. The information explains where to look and how to evaluate fat deposits on slaughter animals.

Pone fat: the fat deposited on either side of the tail. Tail pones are useful in predicting quality grade in all cattle, especially dairy and exotic breeds. Animals showing no fat deposits on either side of the tail head should be considered for the standard quality grade. Select quality grade will show only small amounts of pone fat (about the size of a tennis ball) and choice grade cattle will show moderate amounts of pone fat (about the size of a softball).

Lower round quarter: useful in determining quality grade in all breed types, especially English and English Brahman cross. **Caution:** This is not the cod or udder area but the area inside the rounds. Animals showing standard quality will show no fat. Select quality will show small amounts of fat and choice quality will show a moderate amount of fat.

correct move freer and are more comfortable as they move and stand, figure 8-12. Comfort is a factor in growing and gaining weight because animals that are more comfortable usually grow faster. Refer to the previous chapter in determining structural correctness. Most of the same factors in determining structural correctness in breeding animals apply to market animals as well.

GRADING MARKET BEEF ANIMALS

Market beef animals are graded in two different ways. **Quality grade** refers to the eating quality of beef. **Yield grade** refers to the amount of lean retail cuts that will come from a carcass. Both grades are important in the production, selection, and marketing of beef animals.

DETERMINING QUALITY GRADE

Quality grade is generally determined by two factors: maturity and degree of marbling in the rib eye. **Maturity** refers to age of the animal and is important because the older the animal gets, the tougher and less desirable the meat becomes. The most desirable grades—prime, choice, and select—come from animals that are generally less than 30 months old. The grades of slaughter cattle are as follows:

Prime: the top grade that is most often bought by restaurants to serve as prime rib or prime steaks; the most expensive grade to produce and the most expensive to buy

Choice: the grade most often found in the grocery store and most often bought by consumers

Select: beef from young animals (30 months or less) that does not have the amount of marbling to grade choice or prime; often sold as a healthier grade of beef because of the lower fat content

Standard: beef from young animals that has almost no marbling; often cooked under pressure or using moist cooking methods

Commercial: generally from older animals such as culled cows; the meat may have flavor but may be tough; usually used as ground beef

Utility: this grade can be the same age as commercial grade beef but may not have as much marbling; usually used for ground beef; beef fat may be mixed with the ground lean to give flavor

Cutter: this grade is seldom sold through retail outlets; generally comprised of older thin fleshed animals; usually used for processed meats such as wieners

Because graders have no way of determining for certain the exact age of the animal and the fact that some animals mature at different ages, the physiological age is used. **Chronological age** refers to the actual age of the animal. **Physiological age** of the animal refers to the way the animal has matured as indicated by bone characteristics, ossification of cartilage, and color and texture of the rib eye muscle. As an animal ages, cartilage turns into bone. **Cartilage** is elastic tissues in young animals that turn to bone as the animal gets older and matures. This process is called **ossification**. Ossification begins in the sacral region of the backbone and with advancing age proceeds to the lumbar region and then even later it begins in the thoracic region (buttons) of the carcass. Carcasses can be examined to determine the amount of ossification of the buttons that are at the end of the ribs where they attach to the backbone. Tables 8-1 and 8-2 depict the degree of ossification and the grade of the carcass, respectively.

TABLE 8-1 Maturity Group

VERTEBRAE	A	B	C	D	E
Sacral	Distinct separation	Completely fused	Completely fused	Completely fused	Completely fus
Lumbar	No ossification	Nearly completely ossified	Completely ossified	Completely ossified	Completely ossified
Thoracic	No ossification	Some ossification	Partially ossified	Considerable ossification (outlines of buttons are still visible)	Extensive ossification (outlines of butto are barely visible
Thoracic Buttons ossified	0–10%	10–35%	35–70%	70–90%	>90%

TABLE 8-2 Grade and Marbling Score

QUALITY GRADE	MARBLING SCORE
Prime +	Abundant
Prime °	Moderately Abundant
Prime -	Slightly Abundant
Choice +	Moderate
Choice °	Modest
Choice -	Small
Select +	Slight
Select -	Slight
Standard +	Traces
Standard °	Practically Devoid to Traces
Standard -	Practically Devoid

MARBLING

Once the maturity level of the carcass is determined, the amount of marbling (intermuscular fat) is judged by observing the rib eye area. Usually this is done after the carcass has cooled and begun the aging process. At this time the marbling is easier to see. The final grade is made by estimating the total amount of marbling in the rib eye. Table 8-3 shows the quality grade assigned with each degree of marbling.

QUALITY GRADING LIVE CATTLE

Obviously live cattle are much more difficult to grade than carcasses. Because the evaluator cannot see the fat layers and the marbling, he or she must look for indicators of fat on the live animal. Remember that cattle deposit fat inside the muscle after fat is deposited under the hide and between the muscles. People who grade live animals observe particular portions of the animal's body to determine how

Cod and udder fat: choice pieces appear full; standard pieces show no fat, skin is folded and loose. **Caution:** One must be careful using cod fat, how a steer is castrated may affect his ability to express cod fat. Some steers are cut or clamped so high and tight that there is no place left to deposit fat in the cod region. On the other hand, some are cut or clamped where there is a very large sack left. Choice cod/udder fat appears full; standard cod/udder fat shows no fat, skin is folded and loose.

Brisket: this thickness is useful in determining quality grade in all breed types, especially Exotic, English-Exotic Cross, and Dairy. Standard quality will show no fat in the brisket area and the brisket will appear narrow and pointed. Select quality will show a small amount of fat, while Choice briskets appear full and square.

Flank: depth of flank is a good predictor of marbling especially in English, English cross, and Dairy cattle. One of the best indicators of fatness on an animal is disproportionate depth of body. An animal that has a deeper than natural under line is fat. Standard quality shows no fat, the flank angles up from the elbow to the rear flank. Select quality shows a small amount of fat. Choice quality shows that the under line from the elbow to the rear flank will be approximately parallel to the ground. A fistful of fat may push out in the flank area when the animal walks.

Cheeks and jowl: this can be used to select a quality grade when other indicators leave the evaluator uncertain as to which grade most accurately describes the quality grade (that is, a tie breaker).

Turn over the top: as the amount of fat increases as viewed down over the back of animals. They take on a flat or tabletop appearance. As this condition reaches excessive levels, a shelving effect appears out over the edges of the loin to form a shelf.

Round creases: a good indicator of quality grade in Heifers only. A rope-looking seam of 0.5 to 0.75 inch of fat can be seen extending from the vulva to the udder, between the rounds. These external fat indicators are not absolute, but are useful when other information is not available (e.g., genetics, environment, days on feed, and prior grading information). Using these indicators should increase the degree of accuracy when evaluating groups of slaughter steers and heifers.

YIELD GRADES
Beef Yield Grades

As mentioned, yield grade refers to the percentage of lean, boneless, closely trimmed retail cuts that can be expected from a carcass. Yield grades are denoted by numbers 1 through 5, with yield grade 1 representing the highest cutability or yield of closely trimmed retail cuts. Thus, an increase in cutability means a smaller yield grade number, whereas a decrease in cutability means a larger yield grade number. Usually the most desirable grade is a yield grade 2. If the animal yield grades a 1, the carcass may be too lean and not have enough fat to grade choice. Any yield grade above a 2 may mean that there will be an excess of fat on the carcass and a lot of wastage will occur as the meat is trimmed.

TABLE 8-4 Yield Grade and Percent of Cuts

YIELD GRADE	% OF EXPECTED BONELESS CLOSELY TRIMMED RETAIL CUTS
1	52.3%
2	52.3-50%
3	50-47.7%
4	47.7-45.4%
5	<45%

TABLE 8-5 Factors Influencing Yield Grade

FACTOR	EFFECT OF INCREASE ON YIELD GRADE	APPROXIMATE CHANGE IN EACH FACTOR REQUIRED TO MAKE A FULL YIELD GRADE CHANGE
Thickness of fat over rib eye	Decreases	4/10 inch
% of kidney, pelvic, & heart fat	Decreases	5%
Carcass weight	Decreases	260 lb.
Area of rib eye	Decreases	3 in.

Expected percentage of boneless, closely trimmed retail cuts from beef carcasses within the various yield grades is shown in table 8-4.

Obviously there are several factors that influence the final yield grade. These factors are provided in table 8-5.

Steps in Determining Live Yield Grade

1. The beginning point for calculating a yield grade is the amount of external fat at the twelfth rib. This is an indication of the total amount of fat in the carcass. The thickness of fat is measured at three-fourths the length of the rib eye from the chine (backbone). The preliminary yield grade is determined based on the estimated adjusted fat over the rib eye.

TABLE 8-6 Rib Eye Fat and Yield Grade

FAT OVER THE RIB EYE	PRELIMINARY YIELD GRADE
0.2 inch	2.5
0.4 inch	3.0
0.6 inch	3.5
0.8 inch	4.0
1.0 inch	4.5
1.2 inch	5.0

2. **Kidney, pelvic, and heart (KPH);** fat is evaluated subjectively and is expressed as a percentage of the carcass weight.

TABLE 8-7 Kidney Pelvic and Heart Fat and Yield Grade Adjustments

% KPH	ADJUSTMENT
.5	−.6
1.0	−.5
1.5	−.4
2.0	−.3
2.5	−.2
3.0	−.1
3.5	0
4.0	+.1
4.5	+.2
5.0	+.3

3. Carcass weight; the hot or nonchilled weight in pounds (taken on the slaughter-dressing floor shortly after slaughter). The area of the rib eye is determined by measuring the size (in inches, using a dot grid). The adjustment for area of rib eye is based on the area of rib eye carcass weight relationship in table 8-8. For each square inch by which the area of rib eye is estimated to exceed the area shown for the estimated carcass weight, subtract 0.3 of a grade. For each square inch less than the area shown for the estimated carcass weight, add 0.3 of a grade.

TABLE 8-8 Hot Carcass Weight and Expected Rib Eye Area

HOT CARCASS WEIGHT	AREA OF RIB EYE
350 lb	8.0 sq in
400	8.6
450	9.2
500	9.8
550	10.4
600	11.0
650	11.6
700	12.2
750	12.8
800	13.4
850	14.0
900	14.6

Calculating Yield Grade

Let's calculate a yield grade for the carcass. Below are the measurements taken from the carcass.

Carcass A

0.2 inch of fat over the rib eye

2.5% KPH

551 pound hot carcass weight

12.4 square inch rib eye

Step 1: The carcass has 0.2 inch of fat over the rib eye. From table 8-6 we know that the preliminary yield grade is 2.5.

Step 2: The KPH is 2.5% and from table 8-7 we know that we subtract 0.2 from the preliminary yield grade of 2.5 to equal 2.3.

Step 3: The hot carcass weight is 551 pounds which is rounded to 550 pounds in table 8-8. From this table we know that the average rib eye area for that weight carcass is 10.4 square inches. Since the rib eye area of this carcass is 12.4 inches, we know it exceeds the average by 2 square inches so we need to subtract 0.6 (0.3 × 2). This gives a final yield grade of 1.7 (2.3 - 0.6 = 1.7). The grade is always rounded down to the next whole number, the carcass is a yield grade 1.

Carcass B

0.8 inch of fat over the rib eye

4% KPH

660 pound hot carcass weight

10.8 square inch rib eye

Step 1: The carcass has 0.8 inch of fat over the rib eye. From table 8-6 we know that the preliminary yield grade is 4.

Step 2: The KPH is 4% and from table 8-7 we know that we add 0.1 to the preliminary yield grade of 4. This equals 4.1.

Step 3: The hot carcass weight is 660 pounds which is rounded to 650 pounds in table 8-8. From this table we know that the average rib eye area for that weight carcass is 11.6 square inches. Since the rib eye area of this carcass is 10.8 inches, we know it is below the average (11.6) by 1 square inch so we add 0.3 to the yield grade. This gives a final yield grade of 4.4 (4.1 + 0.3 = 4.4). The grade is always rounded down to the next whole number; the carcass is a yield grade 4.

SUMMARY

The selection of market beef animals is an important component of the beef industry. The type of animal produced must be profitable for the producer and acceptable for the consumer. The correct combination of muscling, size, and finish is essential for both consumer and producer. Too little as well as too much muscling is undesirable and too little or too much finish is also undesirable. Market beef animals should have a quality grade of choice and a yield grade of 2.

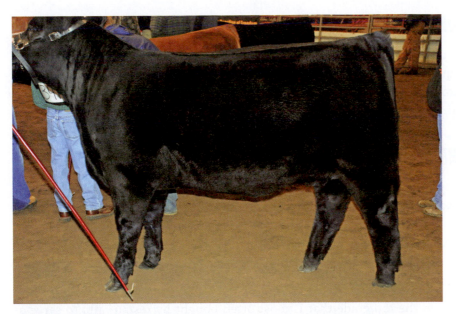

FIGURE 8–10 A properly finished market animal will appear smooth over the ribs. It should also feel spongy when handled. *(Delmar/Cengage Learning. Photo by Ray Herren.)*

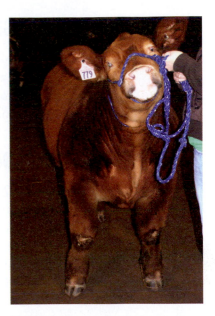

FIGURE 8–11 The brisket on a properly finished steer should appear full but not bulging. *(Delmar/Cengage Learning. Photo by Ray Herren.)*

FIGURE 8–12 This steer is not structurally correct. He stands too straight on his hind legs and is post legged. *(Delmar/Cengage Learning. Photo by Ray Herren.)*

correct move freer and are more comfortable as they move and stand, figure 8-12. Comfort is a factor in growing and gaining weight because animals that are more comfortable usually grow faster. Refer to the previous chapter in determining structural correctness. Most of the same factors in determining structural correctness in breeding animals apply to market animals as well.

GRADING MARKET BEEF ANIMALS

Market beef animals are graded in two different ways. **Quality grade** refers to the eating quality of beef. **Yield grade** refers to the amount of lean retail cuts that will come from a carcass. Both grades are important in the production, selection, and marketing of beef animals.

DETERMINING QUALITY GRADE

Quality grade is generally determined by two factors: maturity and degree of marbling in the rib eye. **Maturity** refers to age of the animal and is important because the older the animal gets, the tougher and less desirable the meat becomes. The most desirable grades—prime, choice, and select—come from animals that are generally less than 30 months old. The grades of slaughter cattle are as follows:

Prime: the top grade that is most often bought by restaurants to serve as prime rib or prime steaks; the most expensive grade to produce and the most expensive to buy

Choice: the grade most often found in the grocery store and most often bought by consumers

Select: beef from young animals (30 months or less) that does not have the amount of marbling to grade choice or prime; often sold as a healthier grade of beef because of the lower fat content

Standard: beef from young animals that has almost no marbling; often cooked under pressure or using moist cooking methods

Commercial: generally from older animals such as culled cows; the meat may have flavor but may be tough; usually used as ground beef

Utility: this grade can be the same age as commercial grade beef but may not have as much marbling; usually used for ground beef; beef fat may be mixed with the ground lean to give flavor

Cutter: this grade is seldom sold through retail outlets; generally comprised of older thin fleshed animals; usually used for processed meats such as wieners

Because graders have no way of determining for certain the exact age of the animal and the fact that some animals mature at different ages, the physiological age is used. **Chronological age** refers to the actual age of the animal. **Physiological age** of the animal refers to the way the animal has matured as indicated by bone characteristics, ossification of cartilage, and color and texture of the rib eye muscle. As an animal ages, cartilage turns into bone. **Cartilage** is elastic tissues in young animals that turn to bone as the animal gets older and matures. This process is called **ossification**. Ossification begins in the sacral region of the backbone and with advancing age proceeds to the lumbar region and then even later it begins in the thoracic region (buttons) of the carcass. Carcasses can be examined to determine the amount of ossification of the buttons that are at the end of the ribs where they attach to the backbone. Tables 8-1 and 8-2 depict the degree of ossification and the grade of the carcass, respectively.

TABLE 8-1 Maturity Group

VERTEBRAE	A	B	C	D	E
Sacral	Distinct separation	Completely fused	Completely fused	Completely fused	Completely fused
Lumbar	No ossification	Nearly completely ossified	Completely ossified	Completely ossified	Completely ossified
Thoracic	No ossification	Some ossification	Partially ossified	Considerable ossification (outlines of buttons are still visible)	Extensive ossification (outlines of buttons are barely visible)
Thoracic Buttons ossified	0–10%	10–35%	35–70%	70–90%	>90%

TABLE 8-2 Grade and Marbling Score

QUALITY GRADE	MARBLING SCORE
Prime +	Abundant
Prime °	Moderately Abundant
Prime -	Slightly Abundant
Choice +	Moderate
Choice °	Modest
Choice -	Small
Select +	Slight
Select -	Slight
Standard +	Traces
Standard °	Practically Devoid to Traces
Standard -	Practically Devoid

MARBLING

Once the maturity level of the carcass is determined, the amount of marbling (intermuscular fat) is judged by observing the rib eye area. Usually this is done after the carcass has cooled and begun the aging process. At this time the marbling is easier to see. The final grade is made by estimating the total amount of marbling in the rib eye. Table 8-3 shows the quality grade assigned with each degree of marbling.

QUALITY GRADING LIVE CATTLE

Obviously live cattle are much more difficult to grade than carcasses. Because the evaluator cannot see the fat layers and the marbling, he or she must look for indicators of fat on the live animal. Remember that cattle deposit fat inside the muscle after fat is deposited under the hide and between the muscles. People who grade live animals observe particular portions of the animal's body to determine how

TABLE 8-3 Shows the Degree of Marbling in each Category.

Degrees of Marbling	Maturity[2]				
	A[3]	B	C	D	E
Slightly Abundant	PRIME				
Moderate			COMMERCIAL		
Modest	CHOICE				
Small					
Slight	SELECT		UTILITY		
Traces					
Practically Devoid	STANDARD			CUTTER	

Table header: **Relationship Between Marbling, Maturity, and Carcass Quality Grade[1]**

[1]Assumes that firmness of lean is comparably developed with the degrees of marbling and that the carcass is not a "dark cutter."
[2]Maturity increases from left to right (A through E).
[3]The A maturity portion of the Figure is the only portion applicable to bullock carcasses.

much fat has been deposited under the skin and between the muscles. A further complication is that different breeds may deposit intermuscular fat with varying amounts of subcutaneous (under the skin) fat. For example, some breeds may not start depositing marbling until they have 0.6 inch of back fat while others may begin depositing marbling at 0.35 inch of back fat.

The following guidelines are used by graders to determine quality grade in live animals. The guidelines are from the USDA publication, *Training Manual for USDA Standards for Grading Slaughter Animals*. The information explains where to look and how to evaluate fat deposits on slaughter animals.

Pone fat: the fat deposited on either side of the tail. Tail pones are useful in predicting quality grade in all cattle, especially dairy and exotic breeds. Animals showing no fat deposits on either side of the tail head should be considered for the standard quality grade. Select quality grade will show only small amounts of pone fat (about the size of a tennis ball) and choice grade cattle will show moderate amounts of pone fat (about the size of a softball).

Lower round quarter: useful in determining quality grade in all breed types, especially English and English Brahman cross. **Caution:** This is not the cod or udder area but the area inside the rounds. Animals showing standard quality will show no fat. Select quality will show small amounts of fat and choice quality will show a moderate amount of fat.

Cod and udder fat: choice pieces appear full; standard pieces show no fat, skin is folded and loose. **Caution:** One must be careful using cod fat, how a steer is castrated may affect his ability to express cod fat. Some steers are cut or clamped so high and tight that there is no place left to deposit fat in the cod region. On the other hand, some are cut or clamped where there is a very large sack left. Choice cod/udder fat appears full; standard cod/udder fat shows no fat, skin is folded and loose.

Brisket: this thickness is useful in determining quality grade in all breed types, especially Exotic, English-Exotic Cross, and Dairy. Standard quality will show no fat in the brisket area and the brisket will appear narrow and pointed. Select quality will show a small amount of fat, while Choice briskets appear full and square.

Flank: depth of flank is a good predictor of marbling especially in English, English cross, and Dairy cattle. One of the best indicators of fatness on an animal is disproportionate depth of body. An animal that has a deeper than natural under line is fat. Standard quality shows no fat, the flank angles up from the elbow to the rear flank. Select quality shows a small amount of fat. Choice quality shows that the under line from the elbow to the rear flank will be approximately parallel to the ground. A fistful of fat may push out in the flank area when the animal walks.

Cheeks and jowl: this can be used to select a quality grade when other indicators leave the evaluator uncertain as to which grade most accurately describes the quality grade (that is, a tie breaker).

Turn over the top: as the amount of fat increases as viewed down over the back of animals. They take on a flat or tabletop appearance. As this condition reaches excessive levels, a shelving effect appears out over the edges of the loin to form a shelf.

Round creases: a good indicator of quality grade in Heifers only. A rope-looking seam of 0.5 to 0.75 inch of fat can be seen extending from the vulva to the udder, between the rounds. These external fat indicators are not absolute, but are useful when other information is not available (e.g., genetics, environment, days on feed, and prior grading information). Using these indicators should increase the degree of accuracy when evaluating groups of slaughter steers and heifers.

YIELD GRADES
Beef Yield Grades

As mentioned, yield grade refers to the percentage of lean, boneless, closely trimmed retail cuts that can be expected from a carcass. Yield grades are denoted by numbers 1 through 5, with yield grade 1 representing the highest cutability or yield of closely trimmed retail cuts. Thus, an increase in cutability means a smaller yield grade number, whereas a decrease in cutability means a larger yield grade number. Usually the most desirable grade is a yield grade 2. If the animal yield grades a 1, the carcass may be too lean and not have enough fat to grade choice. Any yield grade above a 2 may mean that there will be an excess of fat on the carcass and a lot of wastage will occur as the meat is trimmed.

TABLE 8-4 Yield Grade and Percent of Cuts

YIELD GRADE	% OF EXPECTED BONELESS CLOSELY TRIMMED RETAIL CUTS
1	52.3%
2	52.3-50%
3	50-47.7%
4	47.7-45.4%
5	<45%

TABLE 8-5 Factors Influencing Yield Grade

FACTOR	EFFECT OF INCREASE ON YIELD GRADE	APPROXIMATE CHANGE IN EACH FACTOR REQUIRED TO MAKE A FULL YIELD GRADE CHANGE
Thickness of fat over rib eye	Decreases	4/10 inch
% of kidney, pelvic, & heart fat	Decreases	5%
Carcass weight	Decreases	260 lb.
Area of rib eye	Decreases	3 in.

Expected percentage of boneless, closely trimmed retail cuts from beef carcasses within the various yield grades is shown in table 8-4.

Obviously there are several factors that influence the final yield grade. These factors are provided in table 8-5.

Steps in Determining Live Yield Grade

1. The beginning point for calculating a yield grade is the amount of external fat at the twelfth rib. This is an indication of the total amount of fat in the carcass. The thickness of fat is measured at three-fourths the length of the rib eye from the chine (backbone). The preliminary yield grade is determined based on the estimated adjusted fat over the rib eye.

TABLE 8-6 Rib Eye Fat and Yield Grade

FAT OVER THE RIB EYE	PRELIMINARY YIELD GRADE
0.2 inch	2.5
0.4 inch	3.0
0.6 inch	3.5
0.8 inch	4.0
1.0 inch	4.5
1.2 inch	5.0

2. **Kidney, pelvic, and heart (KPH);** fat is evaluated subjectively and is expressed as a percentage of the carcass weight.

TABLE 8-7 Kidney Pelvic and Heart Fat and Yield Grade Adjustments

% KPH	ADJUSTMENT
.5	−.6
1.0	−.5
1.5	−.4
2.0	−.3
2.5	−.2
3.0	−.1
3.5	0
4.0	+.1
4.5	+.2
5.0	+.3

3. Carcass weight; the hot or nonchilled weight in pounds (taken on the slaughter-dressing floor shortly after slaughter). The area of the rib eye is determined by measuring the size (in inches, using a dot grid). The adjustment for area of rib eye is based on the area of rib eye carcass weight relationship in table 8-8. For each square inch by which the area of rib eye is estimated to exceed the area shown for the estimated carcass weight, subtract 0.3 of a grade. For each square inch less than the area shown for the estimated carcass weight, add 0.3 of a grade.

TABLE 8-8 Hot Carcass Weight and Expected Rib Eye Area

HOT CARCASS WEIGHT	AREA OF RIB EYE
350 lb	8.0 sq in
400	8.6
450	9.2
500	9.8
550	10.4
600	11.0
650	11.6
700	12.2
750	12.8
800	13.4
850	14.0
900	14.6

Calculating Yield Grade

Let's calculate a yield grade for the carcass. Below are the measurements taken from the carcass.

Carcass A

0.2 inch of fat over the rib eye

2.5% KPH

551 pound hot carcass weight

12.4 square inch rib eye

Step 1: The carcass has 0.2 inch of fat over the rib eye. From table 8-6 we know that the preliminary yield grade is 2.5.

Step 2: The KPH is 2.5% and from table 8-7 we know that we subtract 0.2 from the preliminary yield grade of 2.5 to equal 2.3.

Step 3: The hot carcass weight is 551 pounds which is rounded to 550 pounds in table 8-8. From this table we know that the average rib eye area for that weight carcass is 10.4 square inches. Since the rib eye area of this carcass is 12.4 inches, we know it exceeds the average by 2 square inches so we need to subtract 0.6 (0.3 × 2). This gives a final yield grade of 1.7 (2.3 - 0.6 = 1.7). The grade is always rounded down to the next whole number, the carcass is a yield grade 1.

Carcass B

0.8 inch of fat over the rib eye

4% KPH

660 pound hot carcass weight

10.8 square inch rib eye

Step 1: The carcass has 0.8 inch of fat over the rib eye. From table 8-6 we know that the preliminary yield grade is 4.

Step 2: The KPH is 4% and from table 8-7 we know that we add 0.1 to the preliminary yield grade of 4. This equals 4.1.

Step 3: The hot carcass weight is 660 pounds which is rounded to 650 pounds in table 8-8. From this table we know that the average rib eye area for that weight carcass is 11.6 square inches. Since the rib eye area of this carcass is 10.8 inches, we know it is below the average (11.6) by 1 square inch so we add 0.3 to the yield grade. This gives a final yield grade of 4.4 (4.1 + 0.3 = 4.4). The grade is always rounded down to the next whole number; the carcass is a yield grade 4.

SUMMARY

The selection of market beef animals is an important component of the beef industry. The type of animal produced must be profitable for the producer and acceptable for the consumer. The correct combination of muscling, size, and finish is essential for both consumer and producer. Too little as well as too much muscling is undesirable and too little or too much finish is also undesirable. Market beef animals should have a quality grade of choice and a yield grade of 2.

STUDENT LEARNING ACTIVITIES

1. Search the internet for images of steers that have won major shows. From the photos, point out desirable characteristics of the animals.
2. Locate some older books on beef production. Using photos in the books, describe how market animals have changed over the years.
3. Visit the meat counter of a large grocery store and look for cuts of beef that are choice and select grades. Try to distinguish the differences in marketing between the two grades.

FILL IN THE BLANKS

1. One factor that has remained fairly constant in selecting market beef animals is consumer _____.
2. Carcasses that have a _____ larger than _____ may be too large for the average consumer.
3. In selecting slaughter steers, preference should be given to the _____ framed steers that finish at _____ pounds.
4. These intermuscular fat deposits, called _____, are what give the meat its _____.
5. Totally lean meat that is devoid of _____ is _____ and rather _____.
6. Selection for extreme muscling leads to the development of cattle with a condition known as

   _____ _____.
7. The hind quarter should be the _____ through the center of the _____.
8. Quality grade refers to the _____ _____ of beef. Yield grade refers to the amount of _____ that will come from a carcass.
9. Physiological age of the animal refers to the way the animal has matured as indicated by _____ _____, _____ of cartilage, and color and texture of the rib eye muscle.
10. KPH stands for _____ _____ and _____ _____.

MULTIPLE CHOICE

1. Tenderness and taste in beef are both related to the:
   a) size of the animal
   b) age of the animal
   c) frame size of the animal
   d) degree of muscling in the carcass

2. Fat is first deposited:
   a) in the body cavity
   b) between the muscles
   c) inside the muscles
   d) under the skin

3. If fed properly, cattle reach the proper amount of finish at about:
   a) 3 years
   b) 2 years
   c) 5 years
   d) 1 year

4. Frame score is determined by measuring the height of the animal at the:
   a) hip
   b) shoulder
   c) head
   d) loin

5. A beef animal that is the widest at the top when viewed from the rear is:
   a) properly muscled
   b) too old
   c) thinly muscled
   d) very young

6. A flat-sided animal with short ribs is:
   a) desirable
   b) lacking in body capacity
   c) the proper hip height
   d) an old animal

7. Quality grade is determined by two factors:
   a) frame size and muscling
   b) amount of back fat and thickness of muscling
   c) bone size and chronological age
   d) maturity and degree of marbling

8. In a properly finished market steer:
   a) the ribs should not be visible
   b) the ribs should be highly visible
   c) observation of the ribs are not important
   d) the ribs should be visible only when the animal is standing

9. Prime beef is sold mostly to:
   a) restaurants
   b) consumers
   c) processors
   d) hamburger chains

10. Cutter grade beef is usually used for:
    a) roasts
    b) hamburgers
    c) steaks
    d) wieners

 DISCUSSION

1. Describe the sequence in which fat is deposited in a beef animal's body.
2. Explain why the proper size of market animal is important.
3. Describe why an extreme amount of muscle is not desirable.
4. Explain how to distinguish between animals that are thick because of fat and animals that are thick because of the proper amount of muscle.
5. Explain why a deep side with well sprung ribs are important.
6. Why is it important that market animals be structurally correct?
7. List in proper order the quality grades of beef.
8. Explain how maturity is determined in a beef carcass.
9. Explain how a live animal that grades choice looks different from a live animal that will grade select.
10. Explain why packers prefer a yield grade 2 over a yield grade 1.

CHAPTER 9
Evaluating Performance Data for Beef Cattle

KEY TERMS

- Performance data
- Index
- Expected progeny difference (EPD)
- Scenario
- Keep cull class

OBJECTIVES

As a result of studying this chapter, students should be able to:

- Explain why performance data are important in cattle selection
- Define the different types of evaluation data
- Discuss how data are used in different scenarios
- Place cattle based on scenarios and performance data

PERFORMANCE DATA

In the preceding chapters, you have learned how to visually evaluate beef animals. This is a valuable skill to have in selecting the proper animals to use in a beef cattle operation. However, visual evaluation is only part of a thorough evaluation process, figure 9-1. The other component is that of evaluation data that has been collected about a particular animal.

There are basically two types of **performance data**. The first is actual data, which are those facts taken from the individual animal. For example, consider an animal's birth weight, the weight of the animal at weaning (205-day weaning weight), the weight of the animal at one year of age (365-day weight), the hip height of the animal at a particular age, or the scrotal circumference of a bull. Another type of actual data is an **index** or

115

FIGURE 9-1 Visual evaluation is only part of a thorough evaluation process. *(Courtesy of American Gelbvieh Association.)*

ratio, which is a measure of how well an animal has performed as compared to animals raised with it. This can be a measure of the genetic differences in the animals because the animals are presumably raised under the same conditions and are treated alike. An index is measured based on a scale of 100, with 100 being the group average. The formula for calculating an index is as follows:

$$\text{Index} = \frac{\text{Individual Animal Weight (or any characteristic)}}{\text{Average Group Weight (or any characteristic)}} \times 100$$

For instance, if a group of calves is weaned when the calves are 205 days old and the average weight of the group is 583 pounds, an index of 100 is equal to 583 pounds. If one of the calves weighs 614 pounds, the calf is said to have a 205-day weaning weight index of 105.3. This means that the animal has outperformed its peers. If a calf has a 205-day weaning index of 82, this would mean that particular calf has performed only 82% as well as the other calves with which it was raised. Other indexes are used for comparing animals on yearling weight, birth weight, and other measures. Additional factors, such as the age of the mother, can be figured into the formula to give a more accurate comparison of the animal's performance, figure 9-2.

Remember, though, that an index is only good for comparing animals within their own groups. Comparing indexes of animals that were raised in separate groups is of little value because the animals almost assuredly were raised under dissimilar conditions. For example, if the index from a heifer raised under range condition where forage is sparse is 103, and another heifer of the same age, raised under very controlled conditions with an abundant source of forage, is 104, then

FIGURE 9–2 Factors such as the age of the mother can also be figured into the formula to give a more accurate comparison of the animal's performance. *(© FredS, 2009. Used under license from Shutterstock.com.)*

FIGURE 9–3 The heifer raised under conditions where feed and care are abundant should have more rapid gains due to environmental conditions and not necessarily genetics. *(© dcwcreations, 2009. Used under license from Shutterstock.com.)*

the indexes are of no use because they were raised under different conditions. This can be said of all actual data because of the differing conditions. The heifer raised under conditions where feed and care are abundant should have more rapid gains due to environmental conditions and not necessarily genetics, figure 9-3. Also, an index from a poorly performing group cannot be compared to an index of an animal from a top performing group. Always keep in mind that indexes are useful only when comparing animals from the same group that were raised together and also the indexes are more useful when more animals are used to calculate the index.

The second type of data is the **expected progeny difference (EPD)**. As discussed in a previous chapter, these data are a compilation of the data collected from the animal's ancestry. The EPD is used to predict the differences that can be expected in the offspring of a particular sire or dam over those of other

FIGURE 9–4 Lower birth weights are desirable because of calving ease. *(© Marcel Mooij, 2009. Used under license from Shutterstock.com.)*

animals used as a reference. The data for calculating EPDs are obtained from the performance data of the progeny from animals of the same breed. For example, an EPD of an Angus bull is calculated using data collected from other Angus cattle. The data are a combination of the performance of his relatives, his actual data, and that of data collected from other Angus bulls.

This is another example of how artificial insemination is useful in the determination of the desirability of a sire because of the tremendous number of offspring made possible by using artificial insemination. This data can be tremendously helpful to a producer who wishes to increase or improve certain traits in the calves produced. For example, if a bull has an EPD of +18 for weaning weight, his offspring can be expected to have a weaning weight of 18 pounds more than the average bull of the same breed. Of course, the bull may have calves that are below average weight, but it should be expected for most of his offspring to have higher weaning weights than average for the breed.

Keep in mind that a larger EPD is not always better. A good example is that of birth weight. If a bull has a birth weight of 83 pounds and his EPD for birth weight is +12, this means that his calves can be expected to average weighing 95 pounds. Obviously a calf weighing 83 pounds should be easier on its mother at birth than a calf weighing 95 pounds, figure 9-4. Carefully consider what the EPD indicates and whether a larger or smaller EPD is more beneficial.

When judging a performance class you will usually be asked to make your final placing based on a combination of the visual placing and placing based on the performance data. At first this may seem like a very complicated task, but in reality it is not all that difficult. Keep in mind that the performance data is important but the best animal according to the data may not be the top-placed animal, figure 9-5.

FIGURE 9–5 This bull may have very good performance data, but he is too straight (post legged) to make him a good breeding bull. *(© Peter Cox, 2009. Used under license from Shutterstock.com.)*

FIGURE 9–6 The type of production means how the animals are raised. This is an important factor in the selection process. *(Courtesy of NRCS, Photo by Jeff Vanuga.)*

For example, a heifer with the top data may have structural defects that may justify moving her toward the bottom of the class. The heifer may stand so straight on her rear legs (post legged) that she will have difficulty walking and living a productive life. You will want to mention this in your oral reasons. A heifer may have the highest weaning weight, the best index for weaning weight, and the top indexing animal in the group but she may look like a steer and appear very unfeminine. This should raise concerns about her femininity and reproductive capability.

When judging a performance class in addition to the production data you will be given a **scenario**. This should consist of three components: the type of production, how the animals are raised, and how the animals are marketed. The type of production means the reason the animals are raised. Are they raised for a terminal market where they will be slaughtered or will they be raised for replacement breeding animals? If they are raised for replacements, are they purebreds or crossbreds? Most times you will be given the production goals. Does the producer want to increase or decrease certain characteristics? For example, the producer may want to increase the weaning weight, calving ease, or postweaning weights. The producer may want to decrease the frame size or the birth weights of calves. These factors should help you determine the type of animals and the EPDs that will help the producer achieve the production goals.

How the animals are raised refers to the conditions used in the operations. Factors such as the environment, available labor, feed, and other inputs available are provided, figure 9-6. Producers who raise animals with plenty of forage, a sufficient level of labor, and a mild climate may want to emphasize growth rate and feed efficiency, whereas a producer who raises cattle on dry range may want to emphasize birth weight and reproductive efficiency, figure 9-7. Following are example scenarios and performance data to evaluate.

FIGURE 9–7 A producer who raises cattle on dry range may want to emphasize birth weight and reproductive efficiency. *(Courtesy of NRCS, Photo by Jeff Vanuga.)*

SCENARIOS
Yearling Bulls

A cattleman runs a large herd of mature Simmental-Hereford cross cows. Futures show that cattle prices for the next several years should remain fairly strong. Because of this, the producer plans to keep all of the calf crop and feed them out to the choice grade in your own farm feedlot. The ranch is located in southern Missouri and the pastures have plenty of high-quality forage. The producer plans to buy an Angus bull for the herd bull. Table 9-1 outlines the performance data.

As you begin to examine the data, pick out the key points in the scenario. Remember that all calves are to be raised for slaughter so the producer is not concerned about the milking potential of the heifers or the birth weight of the bull. Also, because the cows are larger type Simmental cross cows, birth weight should not be a huge concern for the cows to be bred. More concern should be placed on growth rate as indicated by the weaning weight and the yearling weight.

TABLE 9-1 Performance Data of an Angus Bull

BULL NO.	BIRTH WT EPD	WEANING WT EPD	MILK EPD	YEARLING WT EPD
1	+ 1.0	+ 14.0	+ 10.0	+ 18.0
2	+ 3.0	+ 28.0	+ 6.0	+ 31.0
3	+ 4.0	+ 11.0	+ 8.0	+ 14.0
4	+ 2.0	+ 33.0	+ 5.0	+ 39.0
Angus Breed Avg.	3.0	20.0	8.0	30.0

TABLE 9-2 Performance Data of a Purebred Heifer

HEIFER NO.	BIRTH DATE	BIRTH WT EPD	WEANING WT EPD	YEARLING WT EPD	MILK EPD
1	2/16/09	−1.2	+36	+61	+16
2	3/04/09	+4.5	+27	+52	+12
3	3/18/09	+2.2	+14	+33	+14
4	4/16/09	+1.4	+23	+11	+4
Breed Avg EPD=s		+ 0.3	+ 23.5	+ 37.8	+ 8.9

Note that the EPDs for these factors are highest for bulls in rank order 4-2-1-3. The bulls should be placed this way if the placing in the visual evaluation is close to the same.

Purebred Heifers

This scenario differs from that of the yearling bull. This producer raises purebred Herefords and wants to keep all of the heifers for replacements or to sell as breeding heifers. The ranch is in Wyoming and the cattle are produced on semidry range land. There is plenty of hay available and the ranch has labor to monitor and manage the cattle. Rank the heifers as to their desirability as replacement heifers. Table 9-2 displays the data.

Because these heifers are to be used as breeding stock and sufficient labor and feed is available, all of the factors provided should be considered. In analyzing the class, notice that heifer 4 is clearly the bottom-place heifer because she has the lowest EPD for weaning weight, yearling weight, and milking EPD. Number 1 has the highest EPDs in all of the categories except weaning weight which is a positive because of birthing ease. Numbers 2 and 3 logically make up the middle pair, so based on the performance data, the class should be placed 1-2-3-4. As with any performance data, visual evaluation plays an important part in the final placing.

Another type of performance data class is the **keep cull class**. This class consists of eight animals in a group where you visually evaluate the animals based on the performance data. Rather than place them from top to bottom, you are required to pick four from the group to keep as replacements and four to cull from the herd. Both visual and performance data evaluation should use the same criteria as with any of the other classes. Instead of providing a ranking of the animals, you are required to pick only four animals to keep.

SUMMARY

Visual evaluation of cattle has always been used to select animals to be slaughtered for food or to go back into the herd as breeding animals. Only within the past few decades have performance data been an integral component of the selection process. Large amounts of data can be collected because of the widespread use of artificial insemination that allows a single bull to sire hundreds of offspring. Used in combination with visual evaluation, performance data can be a powerful tool in selecting the proper cattle to fit particular production goals.

STUDENT LEARNING ACTIVITIES

1. Create differing production scenarios and tables of production data. Trade your scenarios and data with a classmate. Practice placing classes based on performance data.

2. On the internet, go to www.selectsires.com/beefline_summaries.aspx. Pick out a breed and study the summaries of the sires available. Which of the sires would you select? What criteria did you use?

FILL IN THE BLANKS

1. Actual data which are data taken from the _____ _____.

2. An _____ or _____ is a measure of how well an animal has performed as compared to animals raised with it.

3. An index is measured based on a scale of _____, with _____ being the _____ _____.

4. An index is only good for comparing animals within _____ _____ _____.

5. The data for calculating EPDs are obtained from the _____ _____ of the _____ from animals of the same breed.

6. A heifer with the top data may have _____ _____ that may justify moving her toward the bottom of the class.

7. A scenario should include the type of _____; how the animals are raised; and how the animals are _____.

8. Producers who raise animals with plenty of forage, a sufficient level of labor, and a mild climate, may want to emphasize _____ _____, and _____ _____.

9. A producer who raises cattle on dry range may want to emphasize _____ _____ and _____ _____.

10. Final placings should be based on a combination of _____ evaluation and _____ _____.

MULTIPLE CHOICE

1. Two types of performance data are
   a) structural soundness and indexes
   b) visual and data collected
   c) actual data and indexes
   d) feed consumed and hip height

2. An index is a measure of how well an animal performs compared to animals
   a) in the same group
   b) of different breeds
   c) from different herds
   d) born from the same dam

3. An animal with a weaning index of 108 means that the animal
   a) performed worse than the group average
   b) about the same as the group average
   d) better than the group average
   d) should be culled

4. An index from a low performing group of animals
   a) sets the standard for evaluation
   b) cannot be compared to animals from a high performing group
   c) should be compared to animals in a high performing group
   d) none of the above.

5. EPDs are data collected from
   a) scientific studies
   b) the group average
   c) an animal's ancestry
   d) all of the above

6. When judging a performance class you should make your final placing based on
   a) only the data from the performance records
   b) a combination of the data and visual observation
   c) Mostly visual observation
   d) mostly performance data

7. One type of performance data class is called
   a) cull class
   b) keep class
   c) EPD class
   d) keep – cull class

8. An innovation that has been one of the most useful in compiling performance data is
   a) artificial insemination
   b) selective breeding
   c) visual evaluation
   d) none of the above

9. If a bull has a weaning weight of +20 his offspring can be expected to
   a) weigh 20% more than average
   b) weigh 20 pounds more than the dam
   c) weigh 20 pounds more than the average
   d) weigh 20% more than the sire

10. A high birth weight EPD is
    a) not desirable
    b) very desirable
    c) not very important
    d) only marginally desirable

 DISCUSSION

1. What are some examples of actual data?
2. What is an index or ration?
3. What are Expected progeny Differences (EPDs)?
4. Explain why visual evaluation is important in judging performance classes.
5. What is meant by a scenario?
6. What 3 components are usually included in a scenario?
7. Explain how a keep – cull class works.

SECTION FOUR

Contributed by Michael D. Williams

SHEEP

CHAPTER 10
Selecting Breeding Sheep

KEY TERMS

- Structural soundness
- Reproductive efficiency
- Shoulder
- Capacity
- Leg
- Loin
- Strong top
- Pasterns
- Pigeon foot
- Parrot mouth
- Monkey mouth
- Muscling

OBJECTIVES

As a result of studying this chapter, students should be able to:

- Explain the significance of selecting sheep breeding stock
- Discuss why structural soundness is important in selecting sheep breeding stock
- List feet and leg problems encountered with structurally incorrect sheep
- Know the difference between a structurally correct sheep and one with problems
- Distinguish between a reproductively sound breeding sheep and an unsound breeding sheep

CHARACTERISTICS OF BREEDING SHEEP

Judging breeding sheep is important to the sheep industry because it is the way to improve the characteristics based on selection criteria. Breeding sheep classes are different than market sheep classes due to the reranking of important characteristics. There are six traits to look for when judging breeding stock: (1) balance and style, (2) frame size, (3) soundness and structural correctness, (4) capacity or volume, (5) degree of muscling, and (6) degree of leanness. These six traits are listed in order of importance. In the next chapter, these traits will be explained in greater detail.

These six characteristics are important to consider because the genes that are expressed in the breeding stock will be passed on to their offspring. There are several ways to evaluate breeding sheep. One is the visual evaluation, which is watching and observing the animals. Another is scientific data which has been used extensively due to research conducted by many animal scientists. This data has been collected for many years in order to improve records so that producers can see based on genetics how each generation advances. Evaluators can now rely on this research to figure out which characteristics are most likely to be passed to the next generation.

EVALUATING STRUCTURAL SOUNDNESS IN SHEEP

An important aspect of judging or evaluating sheep is **structural soundness.** Structural soundness refers to the skeletal system and how well the bones support the animal's body. Be aware that the bones making up the skeletal system are the frame upon which the muscles and internal organs are suspended. The bones must support the animal's weight. Bone growth, size, and shape can have quite an effect on the well-being of the animal. Well structured animals are more comfortable as they move around or stand in one place. Structural soundness can also affect **reproductive efficiency**. Rams that have problems moving freely are less likely to be interested in breeding than those that move freely and are comfortable. Soreness, stiffness, and pain in moving greatly reduce reproductive ability.

STRUCTURALLY CORRECT TOP LINES

A skeletal structure that allows a sheep to be comfortable will have a top line that is almost level. The top line refers to the length of the top of the animal's back. This can be seen as the sheep is standing or moving. The top should be strong and the rump should be level. Think of the joints where the legs are attached to the backbone as being hinges that allow the legs to flex and move. The non-level topped sheep has less flexibility and cushion to the joints. As figure 10-1 shows, the scapula and

FIGURE 10–1 The scapula and humerus (shoulder and front leg bone) are more vertical and provide less flex and cushion than a level topped more structurally correct sheep. *(Delmar/Cengage Learning.)*

FIGURE 10–2 A level topped, structurally correct sheep. *(Delmar/Cengage Learning.)*

FIGURE 10–3 The hip and rear leg bones of this sheep are too straight and do not exhibit proper form. *(Courtesy of Michael Williams.)*

humerus (shoulder and front leg bone) are more vertical and provide less flex and cushion than the level topped, more structurally correct sheep in figure 10-2. Note the vertical position of the hip and rear leg bones of the sheep in figure 10-3, as opposed to the same bones that are more nearly parallel to the ground in the sheep in figure 10-4. Again, bones parallel to the ground add more cushion and flex as the animal walks. If these bones are more nearly vertical to the floor, the ends of the bones will jar when the animal walks. This will eventually cause discomfort to

FIGURE 10–4 The proper alignment of the hip and rear leg bones, which is more parallel to the ground. *(Courtesy of Michael Williams.)*

FIGURE 10–5 A sheep that has a steep slope to the pasterns. *(Courtesy of Michael Williams.)*

the animal. On the other hand, if these bones are closer to being parallel, they will act more as a hinge and the animal will be more comfortable walking. The shock of walking will be more absorbed by the ligaments of the joints and not have the jarring effect to the bones. Ligaments are the structures that connect the bones together.

PASTERN STRUCTURE

Pasterns (ankle bones) that are too vertical cause too much jarring as the animal walks. On the other hand, the pasterns should not slope too much because this can cause the sheep to be weak. Weak pasterns are severely discriminated against because they can cause the animal pain and difficulty in walking and standing. Notice the steep slope of the pasterns on the leg of the sheep in figures 10-5 and 10-6.

FIGURE 10–6 Improper pastern structure. *(Courtesy of Michael Williams.)*

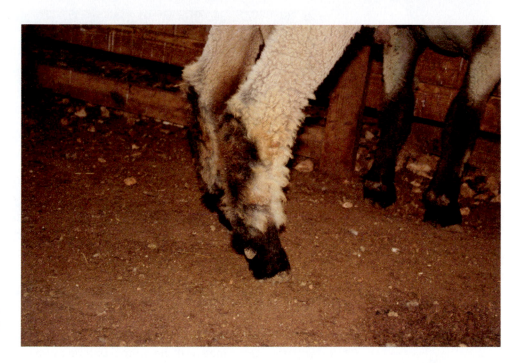

FIGURE 10–7 A sheep with the proper structure of pasterns, which is about a 55 to 60 degree angle to the ground. *(Courtesy of Michael Williams.)*

Pasterns on sheep should be strong, cushioned adequately, and at about a 55 to 60 degree angle to the ground. Figure 10-7 shows the correct structure of the pasterns.

BONE DIAMETER AND FOOT STRUCTURE

Bones should be large in diameter, not small and refined. Common sense says that larger bones are stronger but research has also concluded that they lead to faster growing animals. Larger bones also tend to yield sheep with more capacity

FIGURE 10-8 A sheep with light, small bone structure which is undesirable. *(Courtesy of Michael Williams.)*

FIGURE 10-9 The proper bone structure of a sheep. *(Delmar/ Cengage Learning.)*

for muscle, which will be discussed later. Stronger bones allow the animal to stand longer and walk more comfortably because there is more volume of bone to carry the weight of the animal. Figure 10-8 shows a sheep with light, small bone structure, whereas figure 10-9 shows one with more desirable bone structure.

FEET AND LEG POSITION

All legs of the animal must be sound. This means that from the **shoulder** in the front of the animal to the hip bone in the rear, the legs must be correctly structured. Front legs that are turned out (splayfooted) or turned in (pigeon foot) should be avoided. The rear legs can also point out (cow hocked) or point in (sickle hocked). Figure 10-10 shows a sheep that is splayfooted, whereas figure 10-11 shows a pigeon-foot sheep. Common sense suggests that those animals that walk

FIGURE 10–10 A sheep that exhibits splayfooted feature. *(Courtesy of Michael Williams.)*

FIGURE 10–11 A sheep that exhibits pigeon-footed feature. *(Courtesy of Michael Williams.)*

with these problems cannot do it without some discomfort or pain. It also makes them appear as though they are not walking smoothly, which is an undesirable characteristic. The legs must be in line with the rest of the body to exhibit fluid movement. This is easy to observe in an animal that is either standing or moving. The sheep should move with long, easy strides. Sheep that move with short, choppy steps are either structurally unsound, muscle bound, or both. Structurally sound animals move with a fluidlike grace that appears natural as compared to animals that look like they are struggling to move.

CAPACITY

Capacity, or total volume of the animal, refers to the depth and width of an animal. Preference should be given to sheep that are boldly sprung through the rib cage and wide through the chest floor. They should also have a wide top. The reasoning is that those animals that have greater dimensions in the side, down the top, and through the chest have more space for internal organs such as the heart and lungs. They also have greater capacity for holding feed which leads to higher rates of gain. Breeding animals should be long down the side. Remember that sires and dams pass on their characteristics to their offspring so the longer they are the more meat that they will yield. This will in turn bring more profitability to the producer.

Sheep should stand wide apart at the corners with equal distance between the front and rear legs. This is an indication that they are usually adequately muscled and wider throughout the body cavity. On the other hand, those animals that stand with legs close together are narrower through the body cavity and poorly muscled. Figure 10-12 shows the proper stance of a sheep, whereas figure 10-13 shows a narrow stance.

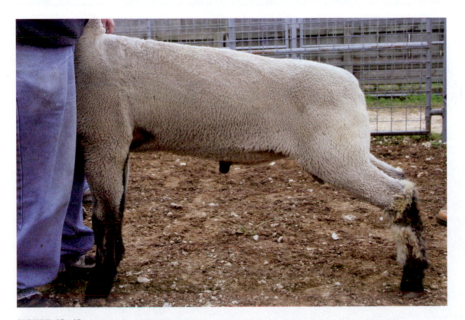

FIGURE 10–12 A sheep with the proper stance desired in the breeding industry. *(Courtesy of Michael Williams.)*

FIGURE 10–13 A sheep that has a narrow appearance. *(Courtesy of Michael Williams.)*

HEAD, NECK, AND SHOULDERS

The head, neck, and shoulders on sheep are also something to consider when selecting breeding stock. The head and the neck should be long and tie into the shoulder at the highest point. The proper angle to the shoulder is critical for good length of stride. The shape and tightness of the front end is also important for proper balance. The point of the shoulder of the lamb in figure 10-14 is not smooth, resulting in the lamb appearing to be heavy and coarse in its shoulder. This lamb is also open shouldered, giving the look of a wide, thick, heavy front end. Figure 10-15 on the left shows a ewe that has the desired tight shape at

FIGURE 10–14 The point of the shoulder of this ram is not smooth, resulting in the ram appearing to be heavy and coarse in its shoulder. *(Courtesy of Michael Williams.)*

FIGURE 10–15 A sheep with the desired tight shape at the top of the shoulder. *(Courtesy of Michael Williams.)*

the top of the shoulder. The lamb in figure 10-16 in the middle picture has the smoothness at the point of the shoulder that is desired, and the shoulder of the lamb on the right in figure 10-17 blends smoothly into the neck and fore rib. The shoulder needs to be smooth because this can have an impact on lambing

FIGURE 10–16 Smoothness at the point of the shoulder is desired. *(Courtesy of Michael Williams.)*

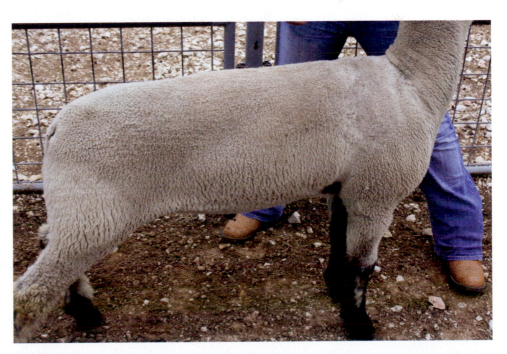

FIGURE 10–17 A sheep that has a shoulder that blends smoothly into the neck and fore rib. *(Courtesy of Michael Williams.)*

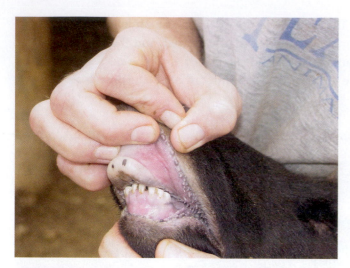

FIGURE 10–18 This lamb has a parrot mouth. *(Courtesy of Dr. Shawn Ramsey.)*

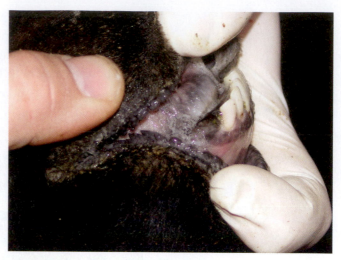

FIGURE 10–19 This lamb has a monkey mouth. *(Courtesy of Dr. Shawn Ramsey.)*

ability. Lambing ability is vital to the efficiency and productiveness of breeding sheep. If the ewe has trouble lambing then help is needed from a human assistant. This reduces profitability and thus should be looked out for when selecting breeding sheep. The mouth should be sound as well with the upper and lower jaws meeting squarely. Some sheep have jaw defects. Jaw defects are highly heritable and severely discriminated against. A sheep could be parrot mouth (where the upper jaw extends past the lower jaw, figure 10-18) or monkey mouth (where the lower jaw extends past the upper jaw, figure 10-19). These problems with the mouth can have a huge impact on how the sheep consumes feed. If the jaw does not meet squarely, then the sheep could waste feed and reduce efficiency and profitability. The sheep also has to work harder and longer to get the same rate of gain. This condition also makes it more difficult for such animals to graze properly because they have a difficult time grasping grass with their mouth.

MUSCLING

Degree of muscling should first be evaluated through the center of the leg for thickness. The second place to examine the lamb is width between the rear feet when it is on the move or standing. It is very important to compare base width, or width at the ground, to top width. In heavy muscled lambs, these should be equal. Be careful to not be tricked by additional thickness due to fat cover. Other areas to evaluate when determining degree of muscling include length of hindsaddle (the loin and the leg), width and length of **loin**, and the shape over the rack (a grooved shape to the rack is desired).

Figures 10-20, 10-21, and 10-22 show lambs that are light muscled, average muscled, and heavy muscled. Note the differences among these lambs in thickness through the center of the leg and base width. The heavy muscled lamb on the far right shows the muscle shape that is desired, being extremely thick through the leg and having a square, wide top shape. A long, wide loin is desirable in market lambs and breeding ewes. Figures 10-23 and 10-24 illustrate the areas to evaluate in determining width and length of loin.

FIGURE 10–20 Light muscled sheep. *(Courtesy of Michael Williams.)*

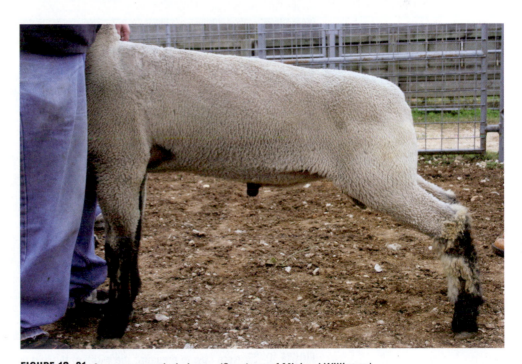

FIGURE 10–21 Average muscled sheep. *(Courtesy of Michael Williams.)*

FIGURE 10–22 Heavy muscled sheep. *(Courtesy of Michael Williams.)*

FIGURE 10–23 Note how to measure the width of the loin on a sheep. *(Courtesy of Michael Williams.)*

FIGURE 10-24 Note how to measure the length of the loin on a sheep. *(Courtesy of Michael Williams.)*

SUMMARY

Visual evaluation of breeding sheep is a vital task for producers. The traits of the parent stock will be passed down to their offspring. Selecting sheep that are structurally sound will help ensure that the animals can move freely and have the best shot at reproducing efficiently. Breeding soundness is just as important. If the animals will not breed and produce young efficiently, they will be of little use and will contribute lower profitability for the producer. While visual evaluation is of importance, it is not the only means to evaluating breeding sheep. The ranking of traits in the order of their importance for judging breeding ewes is as follows:

1. Balance and style
2. Frame size
3. Soundness and structural correctness
4. Capacity or volume
5. Degree of muscling
6. Degree of leanness

STUDENT LEARNING ACTIVITIES

1. Conduct a search of the internet for images of sheep. Try to find examples of sheep that are not structurally desired by producers. Identify the defect by its proper name. Compare your images with others in your class.
2. Contact producers in your area and ask for any advertising images of their breeding stock. If there are none in your area, addresses of producers can be found on the internet.

FILL IN THE BLANKS

1. _____ are the structures that connect the bones together.
2. _____ refers to the width and depth of an animal.
3. A sheep that has its upper jaw extend past the lower jaw is said to have a _____.
4. A sheep that has its front legs pointed out is said to be _____.
5. Degree of muscling should first be evaluated through the _____ of the _____ for thickness.
6. Pasterns on sheep should be strong, cushioned adequately, and at about a _____ degree angle to the ground.
7. _____, _____, and _____ can have quite an effect on the well-being of the animal.
8. The most important part of sheep selection is that of _____ and _____ breeding animals.
9. The proper angle to the _____ is critical for good length of stride.

MULTIPLE CHOICE

1. Which trait characteristic is not important when evaluating breeding sheep?
   a) Balance and style
   b) Degree of muscling
   c) Frame size
   d) Breed type

2. What does structural soundness refer to?
   a) Skeletal system
   b) Digestive system
   c) Reproductive system
   d) Nervous system

3. What do stronger bones allow the animal to do?
   a) Walk tighter
   b) Only stand for a short time
   c) Stand for longer times
   d) Have a hard time standing up

4. Which term describes a sheep with its rear legs pointed out?
   a) Sickle hocked
   b) Splay footed
   c) Pigeon footed
   d) Cow hocked

5. What is another term that can be used for capacity?

   a) Total volume

   b) Length of animal

   c) Feet and leg position

   d) Hindsaddle

6. What does a smooth shoulder have an impact on in breeding ewes?

   a) Feeding ability

   b) Breeding

   c) Lambing ability

   d) Muscling

7. What type of shape is desired over the rack?

   a) Narrow

   b) Wide

   c) Triangular

   d) Grooved

8. Which one of these is a way to evaluate breeding sheep?

   a) Visual

   b) Scientific

   c) Expected Progeny Differences

   d) All of the above

9. What are rams less interested in if structural problems are evident?

   a) Eating

   b) Driving

   c) Breeding

   d) Drinking

10. What does research show that larger bones lead to?

   a) Reduced growth

   b) Rapid growth

   c) Reduced walking ability

   d) Weak pasterns

DISCUSSION

1. What is meant by reproductive efficiency?
2. Explain why animals need to be structurally sound.
3. Why is a level top line so important?
4. List the terms that describe leg defects.
5. Why is it important that an animal have a wide chest and wide ribs?
6. What defect in sheep is highly heritable and should be severely discriminated against?
7. Why are pasterns that are too vertical a problem in breeding animals?
8. What is the most important trait in judging breeding sheep?

CHAPTER 11
Selecting Market Sheep

OBJECTIVES

As a result of studying this chapter, students should be able to:

- Explain the significance of selecting market sheep
- Understand the ranking of traits when evaluating market sheep
- Discuss feet and leg problems of market sheep
- Distinguish between a superior and inferior market sheep
- Know proper terms when speaking about market lambs
- Determine importance of evaluating market sheep

CHARACTERISTICS OF MARKET SHEEP

Market sheep are raised to provide meat for consumers; therefore, they are judged somewhat differently than a breeding sheep class where the main goal is reproduction. When evaluating market sheep, judges should begin by looking at the animal from the ground and then working their way up. From there, judges begin again at the rear and work their way forward. The animals in the class should be ranked based on the importance of the traits they possess. Judges should begin narrowing the class by eliminating any easy placings in the class, then placing the remainder of the class based on the volume and appearance of the important traits.

There are a few evaluation pitfalls a judger must know when looking at market sheep. One of the common issues is selecting lambs that are of an extreme in a particular trait. Single trait selection for any type of livestock can quickly lead to decreased usefulness when other traits are measured. Therefore, lambs should be selected that are well rounded, or complete. These lambs would be above average in all respects, rather than exceptional in one trait and below average in all others. A complete lamb is one that is above average in muscling, of adequate frame size to have economical gains and finish at a market acceptable weight, and is structurally correct.

Evaluating market sheep is different from breeding sheep because of the functionality of the animal. Therefore, the important trait characteristics are arranged differently. The order of ranking characteristics is as follows: (1) muscularity, (2) degree of finish or handling quality, (3) frame size/performance, (4) balance and style, and (5) soundness and structural correctness. A market lamb class is best seen and placed from further distances.

Another common mistake made by many judgers is they tend to get too close to the class, which causes them to not see each individual lamb clearly. There will be time to handle each lamb, but you should already have a placing in mind before handling. Handling just reinforces the original placing and may change one pair but should not change more than one pair. Handling also helps judgers who need to develop speaking points for oral reasons.

MUSCLING

Muscling is the most important trait when evaluating market sheep. Industry has been noted to favor this trait, even in some cases of extreme muscling. Muscling is easier to see from a distance so make sure to back up and get a good look at all the sheep in the class. There are several places to look for muscling. These places are the center of the **leg**, the width between the rear feet when the lamb is on the move and/or standing, the width of chest and forearm, width and shape over **rack** (a grooved shape to the rack is desired), width, length, and depth of **loin**, and length of hindsaddle (the loin and the leg).

Degree of muscling should first be evaluated by observing the center of the leg for thickness. Figure 11-1 shows a lamb with a muscular leg and wide base, whereas figure 11-2 shows a light muscled leg and narrower base. The second place to examine the lamb for muscling is the width between the rear feet when the lamb is on the move and/or standing. It is very important to compare base width, or width at the ground, to top width. In heavy muscled lambs, or the most desirable, these should be equal.

A wide and long loin is desirable in market sheep as shown in figure 11-3. The width of chest and forearm are preliminary indicators of muscle and are advantageous to market lambs. The **hindsaddle** is the back half of the animal. It starts at the last rib and continues to the rear of the animal. This contains the most valuable cuts on lambs. Because of this, the hindsaddle should be equal or greater in length and weight than the foresaddle. The **foresaddle** is comprised of the last rib forward to the base of the neck. The rack is also a fairly high-priced cut in lambs. A grooved shape over the rack would indicate a high degree of muscling.

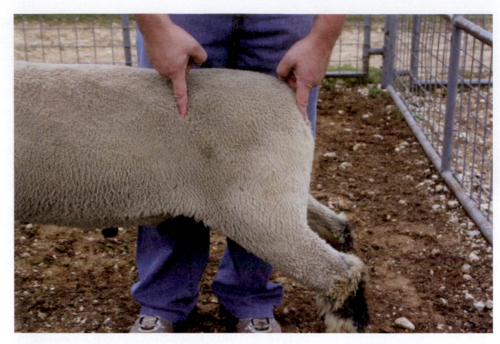

FIGURE 11–1 A sheep with a muscular leg and wide base. *(Courtesy of Michael Williams.)*

FIGURE 11–2 A sheep with a light muscled leg and narrower base. *(Courtesy of Michael Williams.)*

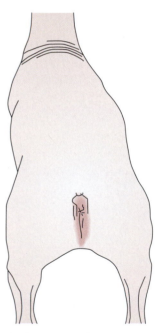

FIGURE 11–3 A sheep with a wide and long loin which is desirable in market sheep. *(Delmar/Cengage Learning.)*

DEGREE OF FINISH OR HANDLING QUALITY

Ideally, lambs should have approximately 0.15 to 0.20 inch of back fat thickness. The desired back fat thickness, or finish, in lambs is much less than that wanted in swine (0.60 inch) and cattle (0.4 inch). Degree of finish is influenced by the

FIGURE 11–4 This lamb shows several signs of too much fat such as a wide top, a poor loose middle, and a heavy front end. *(Courtesy of Michael Williams.)*

amount of muscling, frame size, and stage of maturity. Beware of small framed, light muscled lambs, as they will tend to be too fat. Remember fat sheep will be the widest over their top. Lambs that are lean will be trim over and behind the shoulders. They will also be clean and neat through their under line.

Freshness is the key to **handling quality**. The freshness of a market sheep means that when it is felt over the **shoulder** and down the rack, it will be somewhat firm but not stiff or too tight. Lambs should be fresh directly over their shoulder and over the top of the loin. Stale or nonfresh sheep will feel rigid or sharp through their shoulder and their spine will be felt through the loin. The loin edge must be square, long, and deep.

The lamb in figure 11-4 shows several indicators of too much fat. It has a wide top, a poor loose middle, and a heavy front end. A good rule of thumb to remember when judging market sheep is fat sheep should be placed at the bottom of the class. The lamb in figure 11-5 is very trim by showing a base that is nearly as wide as its top. Fat sheep will normally be widest at the top. The lamb in figure 11-6 shows the proper length and leanness through the foresaddle and hindsaddle. It is very neat and trim through and behind the shoulder as well as being trim and clean all the way through the under line.

FRAME SIZE/PERFORMANCE

Frame size in lambs is used to predict growth potential and to predict size (weight) when properly finished. Usually, lambs should be from average to large in frame size for economical gains. It is also important to remember that frame size varies by breed. Very small and very large frame size lambs should be avoided, as they often finish at weights that are not market acceptable. Market acceptable

FIGURE 11–5 This lamb is very trim by showing a base that is nearly as wide as its top. *(Courtesy of Michael Williams.)*

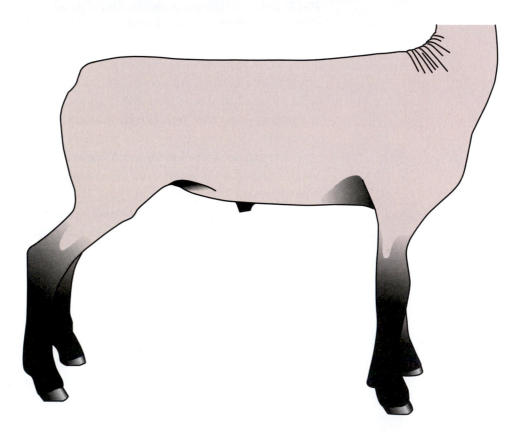

FIGURE 11–6 This lamb shows the proper length and leanness through the foresaddle and hindsaddle. *(Delmar/Cengage Learning.)*

weights are between 95 and 120 pounds. Frame size is often thought of as just height, when actually it includes body length and body **capacity**. Frame size can be compared to a rectangular box, with height, length, and total volume all making an equal contribution. Packers are moving more and more toward the boxing of lambs and do not want excessively fat lambs. Figure 11-7 shows a small framed sheep, whereas figure 11-8 shows a large framed lamb.

FIGURE 11–7 A small framed sheep. *(Delmar/Cengage Learning.)*

FIGURE 11–8 A large framed sheep. *(Delmar/Cengage Learning.)*

BALANCE AND STYLE

When evaluating market sheep for balance and style look at the whole lamb and see how it is put together. The animal's lines should all be clean and flow together. **Balance** or eye appeal is also desirable in your lambs. Eye appeal is a subjective

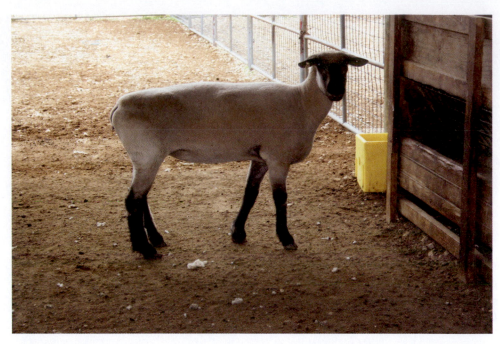

FIGURE 11-9 This sheep is very unbalanced. *(Courtesy of Michael Williams.)*

measurement, but generally lambs should be level in the top line, straight and square over the rump, and not have extreme coarseness through the shoulder, neck, and brisket area. A sheep with balance has equal portions of width, depth, and length with special emphasis being placed on length. For example, if someone were to cut the lamb in half horizontally, both halves should fall to the center. In order to achieve this balance, sheep should be constructed in the form of two Christmas trees. These Christmas tree shapes should be evident when sheep are viewed from behind and from the side, such that they should be wide and deep through the rear (leg) and tight and trim through the front end. Proper balance is important so that the majority of the weight is in the back half of the sheep where the high priced cuts are located. Figure 11-9 shows a lamb that is very unbalanced. This lamb is heavy fronted, deep and low necked, broken topped, too heavy through the middle, and steep rumped. If this example lamb were cut in half, everything would fall toward the front. Compare this with figure 11-10. This illustration shows a sheep that is very balanced and well put together, exhibiting the proper lines of a stylish lamb with the correct Christmas tree shape.

SOUNDNESS AND STRUCTURAL CORRECTNESS

Structural correctness refers to several traits. Your lambs should have sound mouths (not parrot mouths), normal eyes, and be free of any abnormalities. Feet and leg placement should be square with normal width and straightness. Furthermore, lambs should have strong pasterns and no foot problems. When evaluating soundness and structural correctness, start at the ground level and work your way up one joint at a time paying special attention to feet and pasterns, hocks, knees, rumps, and shoulders.

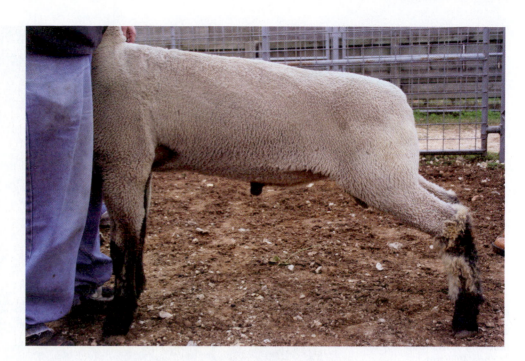

FIGURE 11–10 This sheep is very balanced and well put together, exhibiting the proper lines of a stylish lamb with the correct Christmas tree shape. *(Courtesy of Michael Williams.)*

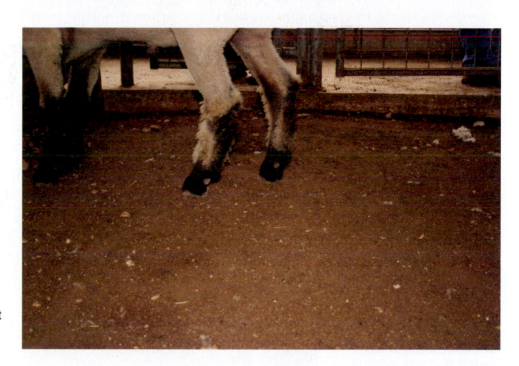

FIGURE 11–11 Pasterns should be strong and have a 55 to 60 degree angle to them so that it acts as a cushion to other joints. *(Courtesy of Michael Williams.)*

Pasterns should be strong and have a 55 to 60 degree angle to them so that it acts as a cushion to other joints, figure 11-11. Figure 11-12 shows an extremely weak pastern that is in danger of breaking down. Sheep should have big feet with even toes that set flat on the surface and square with the animal's body. Figure 11-13 shows a foot with nice, big, even toes. The picture on the right illustrates the pastern and foot **structure** that is desired in sheep. This lamb has the correct set to its pastern, good depth of heel, and it has a big foot that sets flat and even on the ground.

FIGURE 11–12 This extremely weak pastern is in danger of breaking down. *(Courtesy of Michael Williams.)*

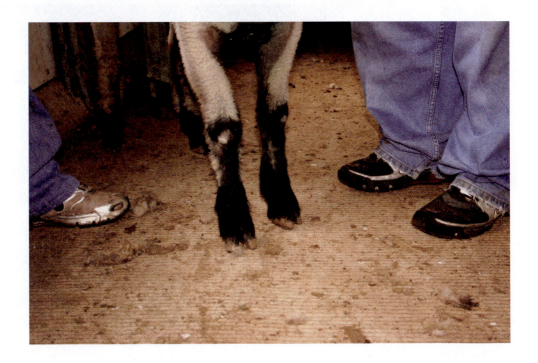

FIGURE 11–13 This sheep's foot has nice, big, even toes. *(Courtesy of Michael Williams.)*

Correct hock structure is critical to the mobility of sheep. They should have a slight angle to them in order to provide the maximum amount of flexibility and power. The bone structure should be clean and free of swelling. Figure 11-14 shows a lamb with the correct set to the hock, whereas figure 11-15 shows an animal that has too much set causing the rear legs to be forced up under the animal. This causes the lamb to have difficulty controlling its rear end. This condition of too much set is called being sickle hocked.

FIGURE 11–14 This illustrates the correct set to the hock. *(Courtesy of Michael Williams.)*

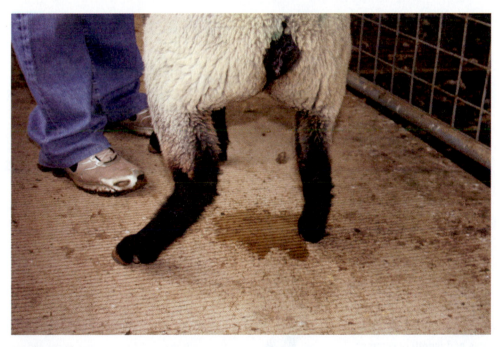

FIGURE 11–15 This sheep has too much set to its hock causing the rear legs to be forced up under the animal. *(Courtesy of Michael Williams.)*

The knees should be square with the body when seen from the front and slightly back when viewed from the side so as to give some cushion when the sheep is in motion. Figure 11-16 shows a sheep that is buck kneed, which is the knees are set too far forward and provides no cushion when moving. Figure 11-17 shows the proper knee structure to provide the ease of motion that is required.

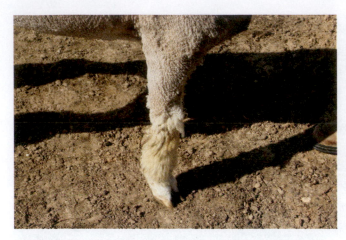

FIGURE 11–16 This sheep is buck kneed, which is where the knees are set too far forward and provides no cushion when moving. *(Courtesy of Michael Williams.)*

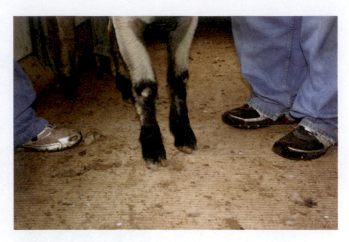

FIGURE 11–17 This sheep has the proper knee structure. *(Courtesy of Michael Williams.)*

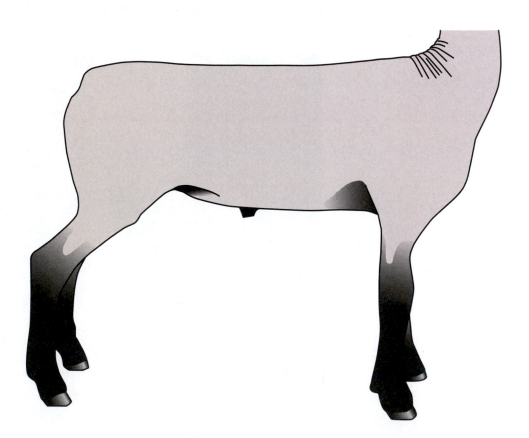

FIGURE 11–18 Note the proper angle and length of rump of this sheep. *(Delmar/Cengage Learning.)*

The **rump** should be average to above average in length and slightly sloping from front to rear. The length is what helps the sheep with the length of stride and capacity for muscle. Figure 11-18 shows the proper angle and length of rump. Short, steep rumped lambs have short strides and do not move freely off the hindquarters.

The shoulders should exhibit the proper angle to them as this affects the length of stride off the front end. The shape and tightness of it also aid in the proper balancing of the lamb. Figure 11-19 shows a desirable shoulder, whereas figure 11-20 shows a course open shouldered, which gives the impression of a wide, thick, heavy front end. This is extremely undesirable.

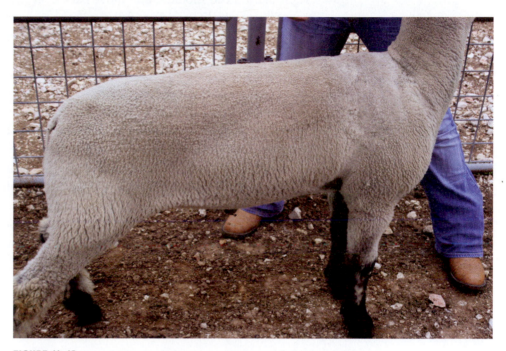

FIGURE 11–19 A desirable shoulder shape. *(Courtesy of Michael Williams.)*

FIGURE 11–20 A coarse, open shoulder. *(Courtesy of Michael Williams.)*

PROPER HANDLING OF MARKET SHEEP

As noted at the beginning of the chapter, before handling a market lamb, the judger should first take a step back and view the lamb from behind. This allows the judger to see the amount of top shape and dimension, how the lamb views out of the hip, how powerful the lamb appears through the leg, and how wide and square the lamb is off the rear skeleton. Next, the judge should handle the lamb right behind the shoulder and move back over the rack and loin. Be sure to measure for muscle dimension and shape in these areas. This is done by looking at the width, length, and depth of the animal. Remember a grooved rack indicates high muscle. Also notice how fresh the sheep appears over the shoulder and loin. Upon that inspection, gently feel the ribs with fingertips to determine fat thickness. Do not rub too hard because eventually anyone can feel the ribs no matter how fat the lamb is. Then take both of your index and middle fingers and place them on the backside of the last rib on each side, and extend both hands until your thumbs are placed on the **hook bones**. This measures the loin length. Next replace your thumbs with your index and middle fingers, and place thumbs on **pin bones**. This measures the length and levelness of the hip. Finally, place both hands under the rear leg as high as possible and touch middle fingers on the inside of the leg. Then wrap your hands around the leg from both directions, and evaluate the distance between the thumbs on the outside of the leg. The greater the distance between the thumbs when the hands are wrapped around the leg, the higher the degree of muscling.

SUMMARY

When evaluating market sheep there are many factors to consider before placing the class. The order of importance is as follows:

1. Degree of muscling
2. Degree of finish
3. Balance and style
4. Frame size
5. Soundness and structural correctness

Muscling is by far the most important trait to consider in market lambs but not the only trait to consider when judging. A superior lamb will exhibit average to above average qualities in all of the traits listed above. Be sure to evaluate the class from a distance first and go ahead and place the class before handling the sheep. Many judges make the mistake of handling first and this causes more confusion than necessary. The traits which are important are best viewed from a distance and handling just reinforces the placing and allows the judger to add to placing reasons. Sometimes after handling, it may be necessary to adjust the placing but be careful not to overanalyze because usually only one pair might need changing. If one feels the need to change more than one pair, step back and reevaluate the class from a distance. Remember judging market sheep should be fun, entertaining, and knowledge building while gaining camaraderie between fellow judgers. Do not forget to relax as well when judging the class.

STUDENT LEARNING ACTIVITIES

1. Conduct a search of the internet for images of sheep. Try to find examples of sheep that are not structurally desired by producers. Identify the defect by its proper name. Compare your images with others in your class.
2. Contact producers in your area and ask for any advertising images of their breeding stock. If there are none in your area, addresses of producers can be found on the internet.

FILL IN THE BLANKS

1. The trait characteristic most important in evaluating market sheep is _____.
2. Lambs should have approximately _____ to _____ inches of back fat thickness.
3. Correct hock structure is critical to the _____ of sheep.
4. A _____ shape over the rack would indicate a high degree of muscling.
5. Degree of muscling should first be evaluated through the _____ of the _____ for thickness.
6. Market acceptable weights are between _____ and _____ pounds.
7. Proper _____ is important so that the majority of the weight is in the back half of the sheep where the high priced cuts are located.
8. Degree of finish is influenced by the amount of muscling, _____, and stage of maturity.
9. Before handling a market lamb, first take a step _____ and view the lamb from behind.
10. The knees should be _____ with the body when seen from the front and slightly back when viewed from the side so as to give some cushion when the sheep is in motion.

MULTIPLE CHOICE

1. What is the first thing you should do before handling a market sheep?
   a) Start feeling the lamb
   b) Look at the lamb over the top
   c) Take a step back and view lamb from behind
   d) Talk nicely to the lamb
2. What is market sheep primarily raised for?
   a) Breeding
   b) Meat
   c) Companion
   d) Style points
3. What do you look for in market sheep when judging?
   a) Extreme characteristics
   b) Below average characteristics
   c) Above average characteristics
   d) Average characteristics

4. What is the proper order of trait characteristics when evaluating market sheep?
   a) Muscularity, degree of finish or handling quality, frame size/performance, balance and style, and soundness and structural correctness
   b) Degree of finish or handling quality, soundness and structural correctness, frame size/performance, muscularity, and balance and style
   c) Balance and style, frame size/performance, muscularity, degree of finish or handling quality, and soundness and structural correctness
   d) Frame size/performance, muscularity, degree of finish or handling quality, balance and style, and soundness and structural correctness

5. What should you already have in mind before handling the class of market sheep?
   a) Nothing
   b) Winning
   c) Losing
   d) Placing

6. What comprises the hindsaddle?
   a) Length of loin
   b) Width of top
   c) Loin and leg
   d) Forerib to rump

7. Where is the second place to look for muscling?
   a) Width between rear feet
   b) Hindsaddle
   c) Center of leg
   d) Rack

8. What are the optimum inches of backfat for a market lamb?
   a) 0.05 to 0.10
   b) 0.15 to 0.20
   c) 0.30 to 0.35
   d) 0.40 to 0.50

9. What can frame size be compared to?
   a) Rectangular box
   b) Circle
   c) Square
   d) Triangle

10. What is the correct angle to the pasterns?
   a) 45–55 degrees
   b) 55–60 degrees
   c) 30–40 degrees
   d) 65–70 degrees

DISCUSSION

1. What is a common mistake made by many livestock judgers when placing a class of market lambs?
2. What is the proper way to handle a market lamb class?
3. What is the condition called where the hock has too much set to it?
4. What portion of the lamb makes up the hindsaddle?
5. What portion of the lamb makes up the foresaddle?
6. Discuss the shape needed in order for the lamb to be properly balanced.
7. What should the rump look like in a sheep that is desirable?
8. Discuss the structural correctness of market lambs.
9. What traits fully encompass frame size?
10. Discuss the areas to evaluate muscling in market sheep.

CHAPTER 12
Wool Judging

OBJECTIVES

As a result of studying this chapter, students should be able to:
- Distinguish between worsted yarn and woolen yarn
- Explain what yield means
- Describe the different types of wool breeds
- Analyze the three major grading systems used to determine fiber diameter
- Describe the characteristics important when evaluating wool
- Explain the wool processing system

KEY TERMS

- Worsted type yarn
- Woolen
- Kemp
- Color
- Character
- Purity
- Staple length
- Crimp
- Uniformity
- Spinning count
- Yield
- Weight
- Hank
- Britch or breech wool
- Lock
- Scouring
- Combing
- Carding

WOOL EVALUATION

Wool judging classes are less subjective in comparison to other judging contests such as livestock judging. Approach each class with the mindset that it is designed to be logical and do not overanalyze it. In wool evaluation, contestants judge two different types of classes; commercial classes and breeding classes. A commercial class is placed on the basis of greatest return to the producer, such as number of pounds of clean wool, whereas a breeding class is placed on the genetic merit of the animals that produced the fleece being judged. For example, what traits are highly heritable and not caused by environmental conditions? There are many different breeds of sheep that are used for wool production, but this chapter will focus on ones used currently in industry. The five types are Rambouillet, Columbia, Corriedale, Debouillet, and Merino.

TYPES OF SHEEP

Rambouillet sheep originated in North Africa in the fourteenth century. Figure 12-1 shows an image of a Rambouillet sheep. American breeders began importing this breed in the mid-1800s and the Rambouillet Association was established in 1889. This breed of sheep is known for its superior, long staple wool. It is also a fine wool breed that is light shrinking and produces a premium price on the market. The average ewe will shear about 10 pounds of wool per year. This wool is mostly used in finer, worsted fabrics and choice scarves.

Columbia sheep were developed by the USDA to replace crossbreeding on the range. Rams of long wool breeds were crossed with high-quality Rambouillet ewes to produce offspring yielding more pounds of wool. Figure 12-2 shows Columbia sheep which are well known for their long **staple length** (3 to 6 inches) and excellent spinning count (54 to 62 spin count). They also produce a high yielding fleece that is bright white and very light in shrinkage.

Corriedale sheep were developed in New Zealand and Australia during the late 1800s and first imported into the United States in 1914. This breed is now distributed worldwide and makes up the greatest population of all sheep in South

FIGURE 12–1 Rambouillet sheep.
(Courtesy of Dr. Richard Coffey.)

FIGURE 12–2 Columbia sheep.
(Courtesy of Dr. Richard Coffey.)

FIGURE 12–3 Corriedale sheep.
(Courtesy of Dr. Richard Coffey.)

America, figure 12-3. A mature ewe fleece weighs between 10 and 17 pounds and yields a high percentage of clean wool (50% to 60%). The staple length on this breed averages between 3.5 and 6 inches.

Debouillet sheep developed in New Mexico in 1920 from Delaine-Merion and Rambouillet crosses. They are well adapted for range conditions in the southwestern United States. They produce a fleece that is high quality and has a deep, fine crimp and high spinning count (62 to 80 spin count). The mature ewe fleece weighs between 10 and 18 pounds with a yield of 35% to 50%. The staple length on this breed is approximately 3 to 5 inches.

FIGURE 12–4 Merino sheep. *(Courtesy of Dr. Shawn Ramsey.)*

Merino sheep are found in extremely high numbers in Australia and were derived from man's first efforts to improve the fiber of its flock in Spain in the fourteenth century. This breed is also know as golden fleece, because it is the finest grading fleece available and sets the standard by which other breeds are measured. The fleece is heavy, soft handling, and has good color. The staple length is about 3.5 inches and the fiber diameter is between 20 and 22 microns. Premium prices are paid by wool buyers for these fleeces and Merino wool is almost exclusively used in high-quality apparel wool. Figure 12-4 shows a Merino sheep.

CHARACTERISTICS OF WOOL

Value of wool is dependent upon its end use, so there are several characteristics to evaluate in order to determine the economic impact of each fleece. The most important characteristics are yield, fiber diameter, and staple length. These are all measurable characteristics. There are more criteria to consider including **character** and **uniformity**, but these traits are more subjective in nature.

Yield

Yield is the percentage of clean wool fibers present in a fleece that has just been sheared. It does not take into account washing and scouring of the fleece. **Scouring** is the actual cleaning of the wool to remove the dirt, grease, and foreign matter. It is usually done in a lukewarm, mildly alkaline solution, followed by clear water rinses, figure 12-5. Yield is vastly variable and manipulated by many factors such as breeding of sheep, soil type, amount or type of range cover, and manner of handling. Due to this, there is an extremely wide spread of yields of wool across the industry. Generally, yield is directly related to fiber diameter so the finer wool fleeces have lower yields, whereas the coarser wool fleeces yield higher percentages. The reason behind this is the finer the wool, the more crimp it has. This tends to lead to capturing of more foreign material in the fleece which reduces the yield during the scouring phase.

FIGURE 12–5 Wool goes through a scouring process. *(Courtesy of Dr. Shawn Ramsey.)*

Fiber Diameter

The next characteristic is fiber diameter, and this is an extremely important property when determining both overall quality and value. Coarser wool has a rapidly decreasing influence on price. In fact, the first U.S. grade standards for wool value used to be based entirely on visual appraisal of fiber diameter. In 1966, the USDA assigned a grade on determining the average diameter and standard deviation of the fiber diameter.

There are three major grading systems used today to determine average fiber diameter: American Blood System, Spinning Count, and Micron Diameter System. The American Blood System is derived from the fine wool of the Merino and Rambouillet breeds and includes: Fine, ½ Blood, ⅜ Blood, ¼ Blood, Low ¼ Blood, Common, and Braid. The **Spinning Count** method refers to the number of hanks of yarn that can be spun from 1 pound of wool top. A **hank** is a length of yarn that measures 560 yards long, figure 12-6. For example, a 70s spinning count will yield 39,200 yards. The Micron Diameter System is the most recent, technical, and accurate measurement of fiber diameter. The micron (1/25,400 of an inch) is used as the measurement. Table 12-1 shows the different systems and what numbers correspond with each.

Staple Length

The average staple length is the third characteristic to consider when evaluating wool, and it determines primarily which system may be used to spin the fibers into yarn. The three classes are staple, French **combing**, and clothing. The selection emphasis on staple length is not overly stressed because excessively long wool provides no technical advantage to the manufacturer. Longer wool is combed into **worsted type yarn** which will be discussed later in the chapter. There is a considerable relationship between staple length of sound wool and average fiber length

FIGURE 12–6 A hank of yarn. *(Courtesy of Michael Williams.)*

TABLE 12-1 Major Grading Systems

AMERICAN BLOOD SYSTEM	SPINNING COUNT	MICRON DIAMETER SYSTEM
Fine	>80s	<17.70
Fine	80s	17.70–19.14
Fine	70s	19.15–20.59
Fine	64s	20.60–22.04
½ Blood	62s	22.05–23.49
½ Blood	60s	23.50–24.94
3/8 Blood	58s	24.95–26.39
3/8 Blood	56s	26.40–27.84
¼ Blood	54s	27.85--29.29
¼ Blood	50s	29.30–30.99
Low ¼ Blood	48s	31.00–32.69
Low ¼ Blood	46s	32.70–34.39
Common and Braid	44s	34.40–36.19
Common and Braid	40s	36.20–38.09
Common and Braid	36s	38.10–40.20
Common and Braid	<36s	>40.20

Ramsey, S. 2008. Course packet, Animal Science 314: Wool and Mohair Evaluation. College Station, TX: Texas A&M University.

in top. Moreover, fiber length in top has an influence on spinning speeds, yarn count, and quality. Table 12-2 shows the staple length requirements by grade.

When discussing staple length, staple strength also must be defined. Sometimes there is a break in the **lock** that is pulled from the fleece. A break is a where the staple pulls apart easily at a specific position at every location within a fleece, figure 12-7. This is caused by environmental influences such as nutrition changes,

FIGURE 12-7 A break in the fleece occurs when the staple pulls apart easily at a specific position at every location within a fleece. *(Courtesy of Michael Williams.)*

TABLE 12-2 Staple Length Requirements by Grade

GRADE	STAPLE	FRENCH COMBING	CLOTHING
64/70s	3.00" and longer	1.25" to 2.75"	Less than 1.25"
60/62s	3.00" and longer	1.50" to 3.00"	Less than 1.50"
56/58s	3.25" and longer	2.25" to 3.25"	Less than 2.25"
50/54s	3.50" and longer	N/A	Less than 3.50"
46/48s	4.00" and longer	N/A	Less than 4.00"
36/40/44s	5.00" and longer	N/A	Less than 5.00"

Ramsey, S. 2008. Course packet, Animal Science 314: Wool and Mohair Evaluation. College Station, TX: Texas A&M University.

severe weather, and disease. When a break is encountered, wool evaluators measure the longest remaining length. Tender wool is also encountered and this is where the wool is weak all over but there is not a definite spot it pulls apart. This is the greatest contributing factor to waste.

Character

Character refers to the eye-catching ability and handling quality of the fleece. This trait can be very difficult to distinguish because it is highly subjective and varies greatly from person to person. Figure 12-8 shows a high character fleece, whereas figure 12-9 shows a low character fleece. The characteristics that can help determine character are color, crimp, and condition.

Color refers to the whiteness and brightness of the fleece. A fleece that appears white and bright as compared to a more yellowish heavy fleece suggests

FIGURE 12-8 A high character fleece. *(Courtesy of Michael Williams.)*

FIGURE 12-9 A low character fleece. *(Courtesy of Michael Williams.)*

greater dying range for the textile industry. Figure 12-10 shows a bright white fleece; figure 12-11 shows a yellow, dingy fleece. This makes the fleece more valuable because it can be dyed to any color. Colored fibers are another issue an evaluator must consider when looking at the color of the fleece. The presence of colored fiber would greatly reduce the value of the fleece because colored fibers and stains are difficult to remove. This makes them very undesirable to the wool manufacturer.

FIGURE 12–10 A bright, white fleece. *(Courtesy of Michael Williams.)*

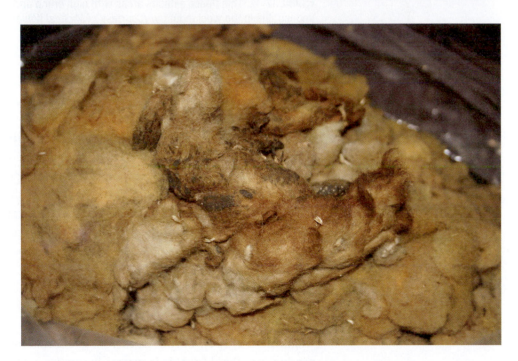

FIGURE 12–11 A yellow, dingy fleece. *(Courtesy of Michael Williams.)*

Crimp and softness correlate to fiber diameter. The more crimp the fleece has, the lower the fiber diameter. This produces a more desirable fleece. An evaluator must also look for consistency with crimp. Some fleeces may exhibit areas with high crimp while other parts have coarse crimp, figure 12-12. This is also undesirable in the industry because this can cause waste material from the fleece.

Condition or **purity** is often associated with color but actually it means free from black or brown fibers, kemp, and hair. **Kemp** are hollow white fibers that will not take dye and therefore discounts the fleece and limits its use in the manufacturing process.

FIGURE 12–12 This fleece exhibits areas with high crimp and areas with coarse crimp which is undesirable in the wool industry. *(Courtesy of Michael Williams.)*

Uniformity

Uniformity is the last trait to look for when evaluating wool. Uniformity is the variation in grade and length of the fleece. When evaluating a fleece for uniformity, gaze at several areas on the fleece and determine grade and staple length. Some areas to consider are shoulder, neck, back, side, britch, and belly. On most fleeces the shoulder, neck, and back tend to be finer crimped and have longer staple length. The belly has fine crimp but usually short staple length, figure 12-13. The **britch**, which is wool from the hindquarters, typically is the coarsest in crimp and often-times looks almost like hair and has a fairly long staple length. All fleeces will have some variation in uniformity, but look for the fleece which has the least amount of variation. That is the one most desirable, and probably at the top of the class.

WOOL PROCESSING

The manufacturing of wool consists of two major divisions. One is apparel and the other is carpet. Carpet is, as you would assume, wool used mainly for carpet production. Wool for this is generally coarser in fiber diameter and fairly long in staple length. Wool carpets last much longer than synthetic ones and are much better on the environment because wool is a renewable resource.

The apparel manufacturing is usually smaller in diameter and is where most of the wool produced in the United States goes. Within apparel wool, there are two manufacturing systems: worsted and woolen. These two types of systems differ because they use different raw materials to process the product. The worsted wool uses wool that has not been previously processed and is used mainly for men's suits. The **woolen** system uses some nonprocessed wool but also contains short fiber by-products from the worsted mills (noils), wool waste, and recycled yarn. These fabrics are used to make softer types such as blankets, overcoats, and women's suits.

FIGURE 12-13 A fleece with belly wool, which is wool that has fine crimp but short staple length. *(Courtesy of Michael Williams.)*

PLACING CLASSES

In evaluating wool classes, the evaluator must first approach the class with the mindset that each class is designed to be logical. Classes typically tend to be less subjective than other contests. First analyze the four fleeces from a distance, which is critical in formulating a general placing. Be sure to take notes on the class and determine distinguishing characteristics of each fleece. Look at the size of the fleeces and give consideration to the larger ones because having pounds of clean wool is most desirable. Handle each fleece to determine **weight** and expected yield. Be sure to pull locks of wool from at least four different areas to evaluate length, character, uniformity, weak or tender fibers, and definite breaks.

There are two different types of classes to evaluate. One is a commercial class and the other is a breeding class. A commercial class is placed on the basis of greatest return to the producer, so areas to emphasize and evaluate critically are pounds of clean wool, grade, staple length, uniformity of grade, fiber strength, character, vegetable matter, and purity. Pounds of clean wool is determined by the combination of grease fleece weight and percent yield. A large, bulky fleece with minimal dirt will yield more pounds of clean wool than a small, compact fleece with an abundance of vegetable matter.

Breeding classes differ slightly from commercial classes because the emphasis switches to placing the class on the genetic merit of the animal that produced the fleece. The major difference between a commercial class and a breeding class is the strength and the vegetable matter content are not taken into consideration, because those two factors are influenced and caused by environmental conditions rather than hereditary. There is also more attention placed on character and purity. The order of traits to look for in a breeding class includes pounds of clean wool, staple length, uniformity of grade, character and purity, and density. Pounds of clean wool and staple length are highly heritable traits and therefore top priority. Uniformity of grade should be indicative of the breed. Character traits are similar to commercial classes except for the addition of distinctness of crimp and if the fiber diameter is representative of the breed. The presence of kemp or hollow fibers is undesirable and density is indicative of clean wool.

SUMMARY

When evaluating wool there are many factors to consider before placing the class. The order of importance is as follows:

1. Pounds of clean wool yielded
2. Grade determination
3. Staple length
4. Character
5. Uniformity

Pounds of clean wool are by far the most important thing to consider when evaluating wool because this is what benefits both the producer and the manufacturer. However, the other traits listed have to be taken into consideration. Remember to evaluate the class from a distance first and then go in to handle the fleeces. When handling them be sure to determine weight and expected yield first. Then start looking at the other traits listed. Also be sure to pull locks from at least four different areas on the fleece to evaluate uniformity.

STUDENT LEARNING ACTIVITIES

1. Contact wool producers in your area and ask if you can come out and observe the shearing process. That way afterward you can evaluate the wool for crimp, character, condition, uniformity, and staple length. If none are in your area ask your teacher for help.
2. Conduct a search of the internet for the wool breeds mentioned in the chapter to familiarize yourself with what they look like.

FILL IN THE BLANKS

1. The Spinning Count method refers to the number of _____ of yarn that can be spun from 1 pound of wool top.
2. Yield is the _____ of clean wool fibers present in a fleece that has just been sheared.
3. The two types of classes judged in a wool contest are _____ and _____.
4. Within apparel wool, the two manufacturing systems are _____ and _____.
5. _____ are hollow white fibers that will not take dye and therefore discounts the fleece and limits its use in the manufacturing process.

6. The major difference between a commercial class and a breeding class are the _____ and the _____ _____ content are not taken into consideration.

7. Be sure to pull locks of wool from at least _____ different areas to evaluate length, character, uniformity, weak or tender fibers, and definite breaks.

8. The _____, which is wool from the hindquarters, typically is the coarsest in crimp and oftentimes looks almost like hair.

9. _____ refers to the eye-catching ability and handling quality of the fleece.

10. The three classes of staple length are _____, _____, and _____.

MULTIPLE CHOICE

1. How is a commercial wool class placed?
   a) Number of pounds of clean wool
   b) Genetic merit of animals
   c) Longest staple length
   d) Highest uniformity

2. Which one of these breeds is considered a wool breed in this chapter?
   a) Hampshire
   b) Southdown
   c) Suffolk
   d) Rambouillet

3. What percentage of clean wool is produced from a Corriedale sheep?
   a) 75–80%
   b) 40–45%
   c) 50–60%
   d) 20–30%

4. What is the most important characteristic when judging commercial wool?
   a) Character
   b) Yield
   c) Condition
   d) Uniformity

5. Which breed is almost exclusively used in high quality apparel wool?
   a) Merino
   b) Debouillet
   c) Rambouillet
   d) Suffolk

6. Which grading system is the most recent and accurate?
   a) American Blood System
   b) Spinning count
   c) Micron Diameter System
   d) Worsted Method

7. How many yards represent a "hank" of wool?
   a) 500
   b) 300
   c) 600
   d) 560

8. What causes a break in the fleece?
   a) Nutrition
   b) weather
   c) disease
   d) all of the above

9. How many different areas on the fleece should locks be pulled from?
   a) 1
   b) 3
   c) 4
   d) 6

10. Where does the britch wool typically come from on the fleece?
    a) Belly
    b) Back
    c) Neck
    d) Hindquarters

DISCUSSION

1. What are the traits to consider when placing a wool class?
2. What is the difference between a break and tender wool?
3. What are the categories of the American Blood System?
4. Name and discuss the five different types of wool breed described in this chapter.
5. What does scouring mean?
6. What are the two major divisions of wool manufacturing?
7. What system of wool manufacturing is used to make men's suits?
8. Discuss the differences between the worsted system and the woolen system.
9. Discuss the staple length requirements by grade.
10. What system is the newest, most accurate for determining fiber diameter?

CHAPTER 13
Evaluating Sheep Performance Data

KEY TERMS

- Adjusted weaning weight
- Adjusted yearling weight
- Fleece quantity and quality
- Grease weight
- Clean weight
- Staple length
- Grade
- Number of lambs born

OBJECTIVES

As a result of studying this chapter, students should be able to:

- Explain why performance data are important in evaluating sheep
- Describe the different production traits used in performance data
- Correctly analyze charts used in performance data
- Use formulas to calculate adjusted weaning and yearling weights
- Describe the EPDs used in sheep performance data
- Analyze data used with specific production scenarios

THE EVALUATION OF SHEEP

Like the other species discussed in this text, sheep are evaluated based on both visual and performance data. In many ways the data collected for sheep are like those for cattle. Both cattle and sheep use expected progeny differences (EPDs) to select superior animals. Remember from the chapter on beef cattle that EPDs refer to the data collected for animals across several herds and include such information as birth date, birth weight, and weaning weight. EPDs refer to how well the animal or animal's offspring is predicted to perform when compared to other animals of that breed. Keep in mind that the use of performance data is to select replacement breeding animals. The use of performance data must be used in conjunction with visual appraisal. Even though an animal may have superior performance data, physical imperfections may render the animal undesirable for use in a breeding program.

Several production traits are used to compile data for judging animals with performance data. Following are the types of data and how they are used.

Birth Date

The birth date refers to the actual day the animal was born. In order to realistically compare animals, the age of the animal must be known. If weights are compared, it is unfair to compare an older animal with a younger one in terms of actual weight. On the other hand, if animals are close in weight and one is older than the other, obviously the younger is the faster growing of the two.

Birth Weight

Newly born animals should be weighed within 24 hours of their birth, figure 13-1. The birth weight is an early indicator of the growthability of the animal. Animals that have an unusually light birth weight may not have the proper weight to get off to a rapid growth or may be unthrifty. Animals with a large birth weight may have problems at birth. The mother may have difficulty birthing an extremely large lamb. Average birth weights are the most desirable because the animal has sufficient body weight to begin to develop properly and is small enough to be easy on the dam at birth.

FIGURE 13–1 Newly born animals should be weighed within 24 hours of their birth. (© 2009 iStockphoto.com/Timothy Large.)

FIGURE 13–2 Ewes may have single lambs, twins, triplets, or on rare occasions quadruplets. (© 2009 iStockphoto.com/Willi Schmitz.)

Number of Lambs

Ewes may have from one to three lambs born at one time. On a rare occasion, they may have as many as four lambs, figure 13-2. Remember that the number born and the number raised are two separate figures. It does little good for a ewe to have triplets if she only raises one of them. From the standpoint of reproductive efficiency, multiple births are more desirable if the ewe can raise all of them. When written as production data, a single birth will be coded "S"; twins will be coded "TW"; triplets "TR"; and quadruplets "Q".

Weaning Weight

The weaning weight of a lamb can be an indication of the growthability, the milking ability of the ewe, or both, figure 13-3. When comparing the weaning weight of lambs there are several factors to consider. For example, a lamb that was raised as a single would probably get more milk than one raised with a twin. Also, an older ewe will probably give more milk than a ewe that is only a year old. A comparison strictly on the basis of actual weaning weigh might be misleading and may not be a good indication of the genetic makeup of the lamb. To help solve this problem, an **adjusted weaning weight** has been calculated to help compare weaning weight factors on a more even basis, table 13-1.

To use table 13-1, assume that you have a wether lamb that was weaned at 90 days and was raised as a twin. The lamb's mother was only one year old and the lamb weighed 10 pounds at birth. The actual weight of the lamb at weaning was 87 pounds. Look in the first column in table 13-1 to locate a wether lamb raised as a twin. Look across that row to the column labeled "1 yr. old" ewe. The number

FIGURE 13–3 The weaning weight of a lamb can be an indication of the growthability, the milking ability of the ewe, or both. *(© Lee O'Dell, 2009. Used under license from Shutterstock.com.)*

TABLE 13-1 Multiply 90-, 120-, or 140-Day Weight By the Appropriate Factor

	AGE OF DAM		
	3 TO 6 YRS. OLD	2 YRS. OLD OR OVER 6 YRS. OLD	1 YR. OLD
Ewe lamb			
Single	1.00	1.09	1.22
Twin–Raised as Twin	1.11	1.20	1.33
Twin–Raised as Single	1.05	1.14	1.28
Triplet–Raised as Triplet	1.22	1.33	1.46
Triplet–Raised as Twin	1.17	1.28	1.42
Triplet–Raised as Single	1.11	1.21	1.36
Wether			
Single	0.97	1.06	1.19
Twin–Raised as Twin	1.08	1.17	1.30
Twin–Raised as Single	1.02	1.11	1.25
Triplet–Raised as Triplet	1.19	1.30	1.43
Triplet–Raised as Twin	1.14	1.25	1.39
Triplet–Raised as Single	1.08	1.18	1.33
Ram Lamb			
Single	0.89	0.98	1.11
Twin–Raised as Twin	1.00	1.09	1.22
Twin–Raised as Single	0.94	1.03	1.17
Triplet–Raised as Triplet	1.11	1.22	1.35
Triplet–Raised as Twin	1.06	1.17	1.31
Triplet–Raised as Single	1.00	1.10	1.25

(1.30) is the adjustment factor. To calculate the adjusted weaning weight, use the following example formula.

Adjusted 90-day weaning weight = 87 pounds × 1.30 = 113.1 pounds

This is the weight that would be used in comparing weaning weights; however, suppose that the lab was not weaned and weighed at precisely 90 days. For an equal comparison, calculate the corrected 90-day weight. The formula is as follows:

$$90\text{-day corrected weight} = \left(\frac{\text{Weaning weight} - \text{birth weight}}{\text{Days of age at weaning}} \times 90\right) + \text{Birth weight}$$

Suppose in our example that the lamb had been weaned and weighed at 85 days old. The calculation for the adjusted weight would be:

$$\frac{87 - 10}{85} \times 90 + 10 = 92.5$$

To figure the 90-day corrected weight, the adjusted weight of 92.5 pounds would be used instead of the 87-pound actual weight in the first example. Also note that some producers wean lambs at 60 days, some at 90 days, and others at 120 days. A lot depends on the production conditions. Usually lambs that are born on range land will be weaned at 120 days. Regardless, the point is that the lambs should be evaluated on an equal basis. If the lambs are weaned at 60 days, use 60 instead of 90 in the formula above. Using the same logic, 120 can be substituted for the 90 also.

Yearling Weight

Just as the weaning weight, the yearling weight is an important indicator of growth in the animal, figure 13-4. In fact, it is a more accurate indication of growth because the weaning weight can be attributed to a large degree by the amount of milk given by the mother. As might be expected, the yearling weight is based on

FIGURE 13–4 The yearling weight is an important indicator of growth in the animal. *(© 2009 iStockphoto.com/David Cannings-Bushell.)*

FIGURE 13–5 Grease weight is the weight of the fleece that has been shorn from the sheep when it is one year old. *(© Margo Harrison, 2009. Used under license from Shutterstock.com.)*

a sheep that is 365 days old. For an **adjusted yearling weight,** simply use the formula for adjusting weaning weight and use 365 days instead of 90 and the age of the sheep in number of days when it was weighed.

Fleece Quantity and Quality

The **fleece quantity and quality** of wool on a sheep is measured at one year of age. There are four measures taken:

   Grease weight is the weight of the fleece that has been shorn from the sheep when it is one year old, figure 13-5. The weight is usually rounded to the nearest tenth of a pound and includes all of the material in the wool before it is washed or scoured.

   Clean weight is the weight of the fleece after it has been cleaned in a wool testing laboratory. Like grease weight, the clean weight is rounded to the nearest tenth of a pound.

   Staple length is the actual length of the wool fiber, figure 13-6. Remember that wool has crimp or waves in the fiber. The length is taken without the fiber being stretched out and is rounded to the nearest tenth of an inch.

   Grade refers to the fineness or diameter of the wool fiber and is measured in microns, figure 13-7. One micron equals 1/25,400 of an inch.

EXPECTED PROGENY DIFFERENCES

As discussed in the performance chapters for the other species, EPDs predict the amount of genetic gain a producer may expect from an animal over the amount expected of the average animal. EPDs are developed for traits that are considered to be economically important. The following sections discuss the EPDs for these traits.

FIGURE 13–6 Staple length is the actual length of the wool fiber. *(© 2009 iStockphoto.com/KCY.)*

FIGURE 13–7 Grade refers to the fineness or diameter of the wool fiber and is measured in microns. *(© Clearviewstock, 2009. Used under license from Shutterstock.com.)*

Maternal EPDs

There are two measures used to predict how good a mother a ewe will be. The **number of lambs born** indicates the ewe's capability to produce more than one lamb at a time. Producers generally look for a positive number. A zero or a negative number indicates poor ability to have twins or triplets. Maternal milk EPDs indicate the amount of milk a ewe may produce, figure 13-8. This is measured by the pounds of lamb weaned. This is the total weight of lambs whether she weans a single or a set of twins or triplets. Maternal EPDs for rams indicate the potential of the ram's daughters.

Growth EPDs

The growth EPDs predict differences in the growth of lambs as measured by weaning, and yearling weights.

Wool EPDs

Wool EPDs indicate the potential for quantity and quality of growth of wool and are calculated for grease fleece weight, fiber length, and fiber diameter.

FIGURE 13–8 Maternal milk EPDs indicate the amount of milk a ewe may produce. *(© 2009 iStockphoto.com/Andrew Hill.)*

TABLE 13-2 EPDs

LAMB #	BIRTH TYPE	BIRTH WT	NO. LAMBS BORN	MATERNAL MILK	WEANING WT	YEARLING WT	DAM'S FLEECE WT
1	S	10.8	−0.38	+0.6	+0.2	−0.43	+2.2
2	TR	11.4	+0.52	+6.0	+3.8	+9.2	+3.4
3	TW	13.4	−0.28	+2.2	+1.3	+1.1	+0.3
4	S	9.3	+0.48	+1.1	+1.4	+2.2	+0.6

SHEEP PRODUCTION SCENARIOS

The performance data in table 13-2 is for Suffolk rams raised in Kentucky on a farm that has adequate resources of feed, labor, and other inputs. The rams will be sold for studs to be used in other purebred Suffolk flocks.

The correct placing for the rams based purely on the performance data is 2-4-3-1. The top-place ram has the highest EPDs for all of the traits and is easily the top-place ram. The number 4 ram was born as a single but the EPD for number of lambs born is the second highest in the class indicating that his daughters will likely produce more than a single. The number 3 ram follows closely in terms of weaning weight and yearling weight EPDs. The number 1 ram is obviously the last-place ram because the EPDs are the lowest in almost all of the categories. Since the rams are Suffolk, the lambs will be raised mostly for meat and the fleece weight is of lesser importance. Of course, as with all performance data, visual evaluation should be combined with performance data evaluation to determine final placing.

SUMMARY

Like all of the other food species of animals, performance data are important components of sheep evaluation. If animals cannot perform based on genetic ability, their visual appearance means very little. Conversely, if the performance data are strong for a particular animal and the animal has physical defects, then sound performance records are of little use as well. Clearly, the best method for evaluating sheep is to use both visual and performance data to choose animals that go into a breed in program.

STUDENT LEARNING ACTIVITIES

1. Conduct an internet search of different breeds of sheep. Determine the type of performance data that is used by the breed association for selecting superior animals. Choose one breed and report to the class.
2. Visit a sheep producer in your area and discuss what types of data the producer uses in selecting replacement animals. Report your findings to the class.
3. Create three different production scenarios and data for each of four animals. Trade scenarios and data with a classmate and practice analyzing the data.

FILL IN THE BLANKS

1. Sheep are evaluated based on both _____ and _____ data.
2. Even though an animal may have superior performance data, _____ _____ may render the animal undesirable for use in a breeding program.
3. Animals with a _____ _____ _____ may have problems at birth.
4. If animals are close in _____ and one is _____ than the other, obviously the younger is the faster growing of the two.
5. When written as production data, a single birth will be coded _____, twins will be coded _____; triplets _____; and quadruplets _____.
6. Some producers wean lambs at _____ days, some at _____ days, and others at _____ days.
7. _____ _____ is the weight of the fleece that has been shorn from the sheep when it is one year old.
8. _____ _____ is the weight of the fleece after it has been cleaned in a wool testing laboratory.
9. _____ _____ is the actual length of the wool fiber.
10. _____ refers to the fineness or diameter of the wool fiber and is measured in _____.

MULTIPLE CHOICE

1. The use of production data is usually to select:
   a) market animals
   b) replacement breeding animals
   c) both market and replacement animals
   d) none of the above

2. Newly born animals should be weighed within:
   a) three days
   b) one hour
   c) one week
   d) 24 hours

3. The most desirable birth weights are:
   a) average
   b) above average
   c) below average
   d) birth weight is of little importance

4. Weaning weight is usually a good indicator of:
   a) good growth genetics
   b) good grazing
   c) the mother's milking ability
   d) the yearling weight of the sire

5. The yearling weight is a good indicator of:
   a) the mother's milking ability
   b) the lamb's ability to grow
   c) the height of the animal at maturity
   d) none of the above

6. Lambs born on a range are weaned at:
   a) 120 days
   b) 60 days
   c) 90 days
   d) 50 days

7. Maternal milk EPDs are measured by:
   a) the pounds of milk the ewe gives per day
   b) the yearling weight of the lambs
   c) the pounds of lamb raised
   d) the genetic makeup of the mother

8. Wool EPDs are calculated for:
   a) grease fleece weight
   b) fiber length
   c) fiber diameter
   d) all of the above

9. EPDs are developed for traits that are:
   a) measurable
   b) of economic importance
   c) easy to visually evaluate
   d) easy to calculate

10. EPDs predict:
   a) the amount of genetic gain expected
   b) the overall size of the animal
   c) what the offspring will look like
   d) none of the above

DISCUSSION

1. Explain why visual evaluation is as important as evaluating performance data.
2. List four types of performance data used to evaluate sheep.
3. Why is it important to adjust actual ages and weights?
4. Distinguish between weaning weight and yearling weight.
5. List four factors considered in determining quantity and quality of wool.
6. Name three types of EPDs used in evaluating performance data for sheep.

SECTION FIVE

Contributed by Kylee Duberstein

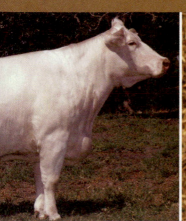   

HORSES

CHAPTER 14
Horse Judging: Anatomy, Terminology, and Rules

OBJECTIVES

By the end of this chapter, you should be able to:

- Analyze the functions of the horse skeletal structure
- Identify different parts of the horse
- Explain the different types of tissue in the horse and how they function coordinately
- Use the proper terms to describe the age and gender of a horse
- Explain why skeletal structure is important in horse selection
- Explain why deviations from ideal structure are not desirable
- Discuss how classes are formed and grouped at horse shows

HORSE EVALUATION

Horses today are used for many purposes, from companionship, to backyard pleasure riding, to competitive performance events. It is important to be able to evaluate a horse based on the job it will be performing. Horse judging involves ranking a group of horses as to their suitability for a specific purpose.

FIGURE 14-1 (A) The horse shows the origination of western horses; this is a horse being used to work cattle. *(© 2009 Roberto Adrian/iStockphoto.)*

FIGURE 14-1 (B) An example of the origin of English riding. This group of horses is participating in a foxhunt, an event where a group of riders pursue a fox over long and rough terrain with natural obstacles using hounds. *(Delmar/Cengage Learning. Photo by Kylee Duberstein.)*

There are many different breeds and types of horses and many different events in which people use horses. A horse **discipline** refers to a particular type or style of riding that a horse may perform. For example, western discipline originated from the days when horses were used to work cattle. Huntseat riding is a particular discipline or type of riding where the horse is evaluated on its ability to carry a rider safely and smoothly over jumps, figure 14-1. Horse **conformation** refers to the physical structure and appearance of the horse. **Halter classes** are classes (competitive events) at a horse show where the judge evaluates a horse's conformation based on the job it will likely be performing. There are halter classes

FIGURE 14–2 Horses in a halter class lined up to be inspected by the judge. *(Delmar/Cengage Learning. Photo by Kylee Duberstein.)*

for most of the different disciplines of riding (western, huntseat, saddleseat, etc.). We will discuss all of these disciplines in greater detail in further chapters. For now, it is important to understand that there are halter classes for different breeds of horses and for the many different disciplines of riding. In these classes, the horse is judged as to whether his conformation will allow him to perform the tasks asked of him in a particular discipline and how well he represents a particular breed or type of horse, figure 14-2.

ANATOMY

A horse's conformation is determined by its skeletal structure. **Anatomy** refers to the structure of the horse. It determines how the horse will look, how the horse will move, what tasks the horse will be able to perform, and how sound the horse will be throughout its life. **Soundness** is a term that describes the horse's physical health and condition. It is generally used to describe whether the horse has injuries that prevent it from being able to function properly. A horse that has such an injury is said to be lame. **Lameness** is a condition that causes the horse pain or reduces the horse's functionality. Such cases generally must be treated by a veterinarian. Some lameness is treatable and the horse is able to return to normal work after a period of time. Some lameness, however, are permanent conditions that will always affect the performance of the horse. Oftentimes, poor conformation will eventually result in lameness in the horse. It is important to learn the anatomy of the horse in order to evaluate its potential ability for a specific discipline. It is also important to learn anatomy to evaluate how likely the horse is to remain sound throughout its lifespan.

THE HORSE SKELETON

The horse's skeleton serves two main purposes. The first is to give shape to the horse and provide the structure to which the muscles, tendons, and ligaments attach. The next function is to protect the vital organs of the horse. For example, the ribs protect the heart, lungs, and other internal organs essential for the horse to live. There are 205 bones in the horse skeleton. The front and hind limbs each contain 20 bones per limb. Bones have three main functions. They primarily act as levers to help in locomotion of the horse. They also are a source of stored minerals that the horse can pull from if needed (particularly the minerals calcium and phosphorus). Thirdly, bones are the site of red blood cell formation. There are four different types of bones. **Long bones** are found in the horse's limbs. These act as levers to aid in locomotion and also are a source of stored minerals. **Short bones** are found in the joints. Their primary function is to absorb concussion and allow the joint to bend. **Flat bones** enclose body cavities and protect vital organs (e.g., the ribs). **Irregular bones** are the vertebrae of the horse which protect the spinal cord.

Soft Tissue

The term *soft tissue* describes the ligaments, tendons, and muscle of the horse. **Ligaments** are the tough, fibrous material that attach bone to bone. These support the bone structure of the horse and also stabilize the joints. **Tendons** are also made of fibrous material and attach muscle to bone. Horses have **flexor tendons** to bend a joint and **extensor tendons** to extend the joint. The horse's muscles contract to pull the tendon and therefore pull the bones of the horse to initiate movement. Muscles are grouped in pairs with one muscle that contracts to extend a joint while one muscle contracts to flex it. Therefore, one muscle must be contracted and the other relaxed to enable the horse to bend a joint. There are many joints, ligaments, and tendons in the horse's legs. It is important to be familiar with these areas since many horse lamenesses occur in the legs. When judging a horse on conformation, unsoundnesses are heavily penalized. You must be familiar with the structural anatomy affected by lameness in order to make good judgments on future soundness.

PARTS OF THE HORSE

In order to understand lameness, you must be able to identify the parts of the leg. Figure 14-3 illustrates a front and hind leg. Terminology that you will commonly hear includes the names of the major joints and bones of the legs. The **knee** is the region on the front leg of the horse between the radius/ulna and the cannon bone. The **hock** joint is in a similar location on the hind leg and is the joint between the tibia and the cannon bone. The **cannon bone** joins the knee or hock joint to the fetlock. The **fetlock joint** connects the cannon bone to the pastern. The **pastern** is composed of three bones: the long pastern, the short pastern, and the coffin bone, figure 14-4. There are specific joints between each of these bones. The **stifle** is similar to the human knee. It is located right above the hind leg and connects the femur to the tibia. The **gaskin** is the large area of muscle on the upper hind leg of the horse (between the hock and stifle).

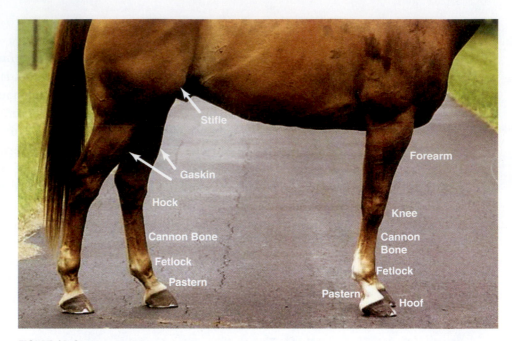

FIGURE 14–3 Important terminology regarding the horse's legs. *(Delmar/Cengage Learning. Photo by Kylee Duberstein.)*

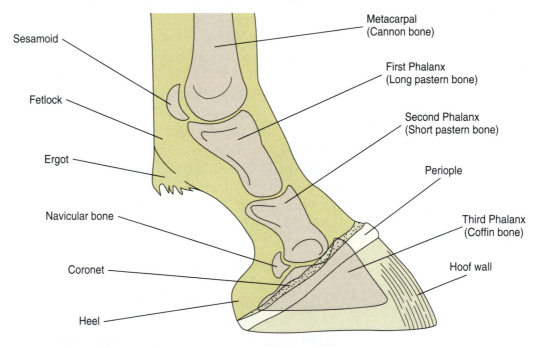

FIGURE 14–4 The structure of the lower leg. The metacarpal bone is the horse's cannon bone, while the first, second, and third phalanx represent the long and short pastern and the coffin bone. *(Delmar/Cengage Learning.)*

Before you can properly discuss and judge a horse's conformation, you must be able to recognize the parts of the horse and the terminology that describes them. Figure 14-5 shows the parts of the horse commonly referred to when evaluating conformation. This section will focus on defining terms relating to the parts of the horse. Continue to refer to Figure 14-5.

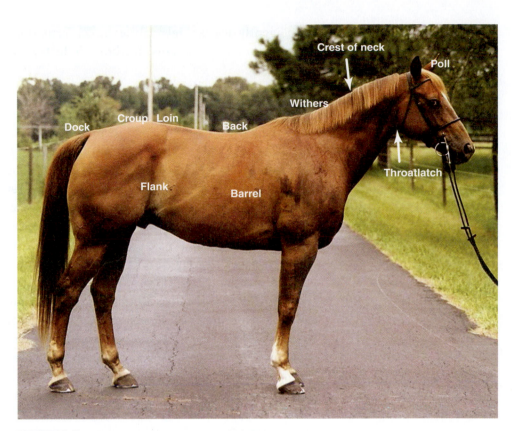

FIGURE 14–5 Important terminology regarding the horse's body. *(Delmar/Cengage Learning. Photo by Kylee Duberstein.)*

Toward the head of the horse, terms that are commonly mentioned include the poll, the throatlatch, and the crest of the neck. The **poll** is the area directly between and behind the ears. It is the poll joint at the beginning of the neck that allows the horse to move its head. The **throatlatch** is the area where the windpipe on the underside of the horse's neck meets the head. Typically, bridles and halters have straps that run under the throatlatch. The **crest** is the top portion of the neck where the horse's mane grows. The back of the neck attaches to the horse's **withers**, which is the tallest part of the thoracic vertebrae. This is the area directly above the shoulder blades. The height of the horse is measured from the ground to the top of the withers. Every 4 inches equals one **hand**. Therefore, a horse that is 16.2 hands high is 66 inches tall (16 × 4 + 2 inches). The horse's **shoulder** refers to the scapula and associated muscles. It is the area right below the withers. The shoulder runs from the withers to the point of the shoulder. Right behind the withers is the horse's back. The back reaches to the last thoracic vertebrae. The back is where the saddle sits. Directly behind the back is the horse's **loin.** This is the area from the last rib to where the croup starts. The **croup** is the region from the beginning of the hip to the dock of the tail. It is considered the **top line** of the horse's hindquarters. The **dock** is the point where the horse's tail connects to its hindquarters. The horse's **barrel** refers to the large main body of the horse where the ribs enclose the vital organs. The horse's **hindquarters** refer to the large muscular area of the horse that is behind the barrel. The **flank** is the area where the hindquarters meet the barrel. Understanding these main terms will allow you to intelligently discuss and analyze the conformation of the horse with other judges, competitors, veterinarians, and horse industry professionals.

AGE AND GENDER

Other important terminology to understand relates to age and gender of the horse. A horse starts out life as a **foal** or **suckling foal** and is generally referred to as that until around six months of age. At that point, the horse is weaned from its mother and is referred to as a **weanling** for the remainder of its first year of life. Horses traditionally age on January 1 of each year. This means that January 1 of each year is typically considered every horse's birthday. By doing this, horse shows have a way to group horses by age for specific events (e.g., two-year-old classes, three-year-old classes). In disciplines where there are important age restricted classes, this method has caused a push in breeders to breed horses early in the year. By doing so, the foals are born as close to January 1 as possible and are therefore more mature than those born later in the year. Horses are typically referred to as **yearlings** when they are in their one-year-old class. **Mature** or **aged horses** are typically five years old and older. Many different disciplines have **futurities**, or higher money stakes classes, for young horses.

Terminology relating to gender and breeding is important to understand since classes are often grouped by gender. In young horses, intact male horses are referred to as **colts** and females are referred to as **fillies** until they are around five years of age. At five years, intact males are referred to as **stallions** and females are called **mares**. A castrated male is referred to as a **gelding**. A stallion used for breeding is referred to as a **stud**. Halter classes will often be grouped by age and gender (e.g., two-year-old fillies). This enables a judge to compare the horse to a group of peers. Horses have different physical attributes relating to their age and gender. Therefore, it is difficult to compare horses of dissimilar groups (e.g., an aged stallion to a yearling filly). Another area of great interest when selecting horses is their pedigree. When discussing how a horse is bred, a horse is described to be out of a mare (its mother) and by a stallion (its father). When analyzing a horse's bloodlines, the product of a mare describes offspring out of the same mare while the get of a sire refers to offspring sired by the same stallion.

HORSE JUDGING CONTESTS

When judging horses, it is important to be able to organize your thoughts, correctly rank the horses, and defend your decision with a logical set of reasons. A **horse judging contest** is a competition in which a person is scored on how accurately he or she places a class and how well he or she defends the reasons for placing the class a certain way. A typical horse judging contest has eight to twelve classes with four horses in each class. The classes will be a combination of halter and performance classes. The class is scored based on two things: accurate placing of each class and the organization/presentation of **oral reasons**. Giving oral reasons involves the presentation of a short oral defense for the way you placed a certain class. This is done from memory and is two minutes or less in time.

Placing Horses and Giving Reasons

To place a halter or performance class, you must rank the horses from top to bottom. It is usually best to divide the horses into pairs and have a top pair, a middle pair, and a bottom pair. Therefore, in a class of four horses, your three pairs would be your first- and second-place horses, second- and third-place horses, and third- and fourth-place horses. This gives you the best opportunity to organize your reasons for why you placed the class as you did. A maximum of 50 points is awarded for correct placing of a class. The **cut system** is typically used to determine scores. If a contestant places the horses exactly as the judge does, no points are deducted. If not, the amount deducted for having a placing different from the judge is based upon the relative difference in the animals that are being judged. Therefore, if two animals are very similar, the point deduction will not be as large as if there is a very clear distinction between the horses.

Cuts are assigned to each pair of horses (1 vs. 2, 2 vs. 3, and 3 vs. 4). Small cuts indicate that the two animals being compared are very similar, whereas large cuts indicate more obvious differences between the animals. Cuts have a point value of 1 to 7 and have been defined by the National Collegiate Horse Judging Coaches Association, Table 14-1. The sum of the three cuts (top, middle, and bottom pair) cannot exceed 15. When tallying a contestant's **placing score,** six comparisons must be made between the contestant's results and the judges' results: 1 vs. 2, 1 vs. 3, 1 vs. 4, 2 vs. 3, 2 vs. 4, 3 vs. 4. When looking at each, if the contestant's comparison is the same as the judges', then there is no point deduction. If a comparison differs from the judge's, a point deduction is made. If the comparison can be switched to be the same as the judge's by simply reversing what the contestant has

TABLE 14-1 This table shows the description of cuts as defined by the National Collegiate Horse Judging Coaches Association.

CUT	DESCRIPTION
1	Horses are extremely similar with no obvious argument why one is placed over the other. Placings will be strictly a matter of personal preference. Officials will vary in their placings.
2	Horses are very close, but an argument can be made for 1 based on one or two advantages. The majority of officials will agree on the placing, while half the contestants could logically switch the pair.
3	Horses are of similar quality, but a strong argument can be made for placing one over the other. Logical placing is based on one animal possessing several distinct advantages and/or one animal exhibiting several faults. All officials will agree and approximately 2/3 of the contestants will find the placing.
4	Horses are not of similar quality and switches cannot be logically argued. All officials and 90% of contestants would correctly see the placing.
5	Horses express extreme differences and placing is obvious to everyone on first quick observation. Pair consists of an inferior animal and a consistent winner. Only the inexperienced would miss placing.
6	Horses are not even comparable. Pair would consist of a champion caliber animal/performance and an animal/performance not of show quality.
7	Largest cut. Horses are worlds apart. Pair would consist of world champion caliber and non-show quality caliber, or animal which is disqualified for lameness.

written, then the deduction equals the point value of the cut. For example, when comparing horse 1 and 2, if the contestant has horse 2 placed right above horse 1, then the deduction would equal the point value of the cut assigned to the top pair. If the comparison cannot be changed by simply switching the placing, the deduction will equal the sum of more than one cut. For example, if a judge placed a class 2-1-4-3 and a contestant placed the same class 3-4-2-1, then when comparing the first-place horse to the second-place horse, there would be no deduction (the judge placed horse 2 over horse 1 and so did the contestant, although not at the top of the class). However, when comparing the first-place horse to the third-place horse, the judge placed horse 2 over horse 4 while the contestant placed horse 4 over horse 2. The deduction would be equal to the sum of the cut between horse 2 and 1 and between horse 1 and 4. The total score is equal to 50 minus the sum of all of the deductions.

ORAL REASONS

A maximum of 50 points is awarded for oral reasons. Giving oral reasons involves verbally presenting a defense for the way you placed a certain class. Oral reasons are scored in five areas: organization, relevancy, accuracy, terminology, and presentation. Organization scores are given based on the format of the set of reasons. Reasons should flow in a logical, easy to understand sequence. The easiest way to organize a set of reasons is to use the top pair, middle pair, bottom pair organization and compare animals in each pair. Relevancy refers to describing the points that are most important. Oral reasons cannot exceed two minutes in length. Therefore, it is important to state the most important reasons for placing one horse above another in a pair.`

There may be several reasons for placing one horse above another but if time is a factor, choose the most important differences. Accuracy is a critical part of oral reasons. Only list things that you are sure of; do not speculate on what might happen to the horse in the future or what the horse might or might not be good at. Stick to what you can see and describe. Terminology describes the words you use to paint a picture, so to speak. This is what sets one contestant apart from another. Using correct terminology shows the scorer that you are knowledgeable. Finally, presentation skills are critical in scoring oral reasons. This includes both verbal and nonverbal presentation. It is important to display poise and confidence in your decisions. You must emphasize the key points that you want the scorer to remember. Proper dress is important; always wear clean, professional attire to judge a horse show or participate in a horse judging competition.

In general, an oral reasons score of 46 to 50 points is good to excellent, a score of 41 to 45 points is above average to good, a score of 36 to 40 points is average, and a score of 31 to 35 points is below average. Scores lower than 25 points are not usually used if the contestant has decent presentation and organizational skills. Scores below 25 points may be given if the contestant misses a very obvious problem, such as a clear lameness in one of the horses.

Oral Reasons Structure

Oral reasons typically begin with an opening statement. This involves stating the class name and the way you placed the class. For example: "I placed this class of yearling fillies 3-2-1-4." You then move into comparisons of the horses to each other. Start with your top pair and describe why you placed the first-place horse over the second-place horse. It is good to have a **grant statement** in your reasons. This is a statement where you grant that the second-place horse is superior to the first-place horse in a certain area *but* the first-place horse has more superior qualities that still allow him to remain in first place. This does not discredit your placings; it shows that you know that the first-place horse is not superior in every way but it does have enough attributes to allow it to be ranked first in this group of horses. Now you move on to the middle pair of horses and compare the second-place horse to the third-place horse. Again, first list the areas in which you thought the second-place horse was superior to the third-place horse. Then grant that the third-place horse was superior to the second-place horse in particular areas but the second-place horse had enough attributes to place it second. Finally, describe the bottom pair and discuss why you placed the third-place horse over the last-place horse. Grant that the last-place horse had attributes that were superior but then list the faults that it had which forced it to the bottom of the class. Finish with a closing statement that reiterates the class name and the way you placed the class. For example: "For these reasons, I placed this class of yearling fillies 3-2-1-4."

SUMMARY

Horse judging involves selecting horses that are best suited for the purpose they are performing or will be asked to perform in the future. Halter classes are judged events that involve evaluating a horse's conformation and ranking it in relation to a group of its peers. In order to be able to evaluate horse conformation, you must first understand the anatomy of the horse and the terminology used to describe the parts of the horse as well as its age and gender. Competitive horse judging involves judging a set of eight to twelve classes. These classes typically consist of four horses per class and are a mixture of halter and performance classes. Horse judging contestants are scored both on the accuracy of their placement of the class and on the oral reasons they present to defend their decisions.

STUDENT LEARNING ACTIVITIES

1. Look at either a real horse or a diagram of a horse and label all parts of the horse's body and legs.
2. Find an encyclopedia or book on horse anatomy and research at least one tendon and one ligament found in the horse's body and what lameness might affect each.
3. Pick a certain registered horse and describe its age, gender, and pedigree using terminology described in this chapter.

TRUE OR FALSE

T   F   1. The ribs are flat bones.

T   F   2. The femur is a short bone.

T   F   3. There are two bones that comprise the pastern.

T   F   4. The dock is found on top of the horse's head between the ears.

T   F   5. Ligaments attach muscle to bone.

T   F   6. Geldings are castrated male horses.

T   F   7. A seven-year-old horse would be considered mature.

T   F   8. Oral reasons are scored in five areas: organization, relevancy, accuracy, terminology, and presentation.

T   F   9. The cut system is used to score the contestants' placings in a horse judging competition.

T   F   10. A grant statement tells why you placed one horse above another.

FILL IN THE BLANKS

1. _____ refers to the physical structure and appearance of the horse.

2. _____ are classes (competitive events) at a horse show where the judge evaluates a horse's conformation based on the job it will likely be performing.

3. A horse _____ refers to a particular type or style of riding that a horse may perform.

4. _____ are conditions that cause the horse pain or reduce the horse's functionality.

5. Tendons attach _____ to _____.

6. Horses have _____ to bend a joint and _____ to extend the joint.

7. The _____ is similar to the human knee.

8. The _____ is the area where the hindquarters meet the barrel.

9. Horses are typically referred to as _____ when they are in their one-year-old class.

10. The _____ describes offspring out of the same mare.

DISCUSSION

1. Where would you find long bones, short bones, flat bones, and irregular bones?

2. Explain the terms *soundness* and *lameness* as they pertain to the horse.

3. Why are a horse's leg muscles grouped in pairs?

4. How and where do you measure a horse's height?

5. What does the "get of a sire" refer to?

6. What is a horse judging competition?

7. What two things are scored in a horse judging competition?

8. Describe a format for giving oral reasons.

9. Describe what the cut system is and how it is used for scoring horse judging competitions.

10. What five criteria are used to score oral reasons?

CHAPTER 15

Judging the Halter Class— Evaluating Conformation

KEY TERMS

- Balance
- Structural correctness
- Type
- Muscling
- Quality
- Blemish
- Splint
- Windpuff
- Ringbone
- Sidebone
- Bowed tendons
- Capped hock
- Curb
- Bog spavin

OBJECTIVES

As a result of studying this chapter, students should be able to:

- Explain why conformation is important to performance and soundness
- Describe leg anatomy and structural deviations associated with front and hind legs
- Describe how structural deviations affect a horse's movement
- Describe the correct ratios and proportions of the horse's body parts
- Analyze why halter judging is an important part of the horse industry
- Describe common blemishes and lamenesses associated with horses
- Discuss why the ratios and proportions of the horse's body parts influence how it will perform
- List the areas to evaluate when judging a halter class

BASICS OF THE HALTER CLASS

Halter classes are popular events at both open shows and breed restricted shows. In these classes, the horse is presented to the judge without a saddle on, and the judge will evaluate the horse based on its conformation. The judge is looking for structural correctness, suitability for the desired discipline, and how well the individual represents the breed or type.

Typically, there are two main groups of halter classes: stock horse and hunter in hand. Stock horse classes are judged on the horse's suitability for western performance events, such as western pleasure, reining, and cutting. These horses are traditionally shorter than hunter horses and tend to be more heavily muscled. They originated from the days when horses were used to work cattle. Hunter horses tend to be taller with a longer, leaner muscle pattern and a longer stride. This type of class originated from horses that went on foxhunts and needed a long, flowing stride and good jumping form.

It is important to understand that in a halter class, the horse is not actually performing any task related to its discipline. It is simply being judged on its conformation and resemblance to breed standards. The horse's conformation is generally judged from all sides as the horse is presented standing before the judge. In some classes, the horse will be asked to walk and trot so that the judge can form an opinion as to how the horse moves. It is important to keep in mind that although different breeds will have different characteristics, structural correctness is similar for all horses.

CRITERIA FOR JUDGING THE HALTER HORSE

When evaluating a horse's conformation, there are several important things to consider. The five criteria by which horses are typically judged are balance, structural correctness, type, muscling, and refinement. **Balance** is one of the more critical aspects to evaluate when examining the horse. To move and perform well, the horse needs to be balanced. Balance refers to equal distribution of muscling and weight from the front of the horse to the back of the horse, from its top to its bottom, and from side to side, figure 15-1. Proper balance enables the horse to carry itself in a manner to allow for easy maneuverability, greater power, and smoother movement. Balance is determined primarily by the skeletal structure of the horse, although muscling also plays a role.

Structural correctness is critical in conformation classes and includes proper structure of bone and also proper angles and ratios of different parts of the body. Structural correctness is tied very closely to balance. Structural correctness and balance also influence the way a horse moves. Oftentimes, when judging conformation, horses are asked to walk and trot to determine if their quality of movement would be suitable for a good performance horse. Structural correctness of the legs and proper proportions of the body are the determinants of a horse's movement.

The term **type** in horses refers to how well that horse represents its particular breed. Most breeds have unique qualities by which they can be identified. Judging a horse on its type refers to judging how well it resembles the ideal horse of that breed.

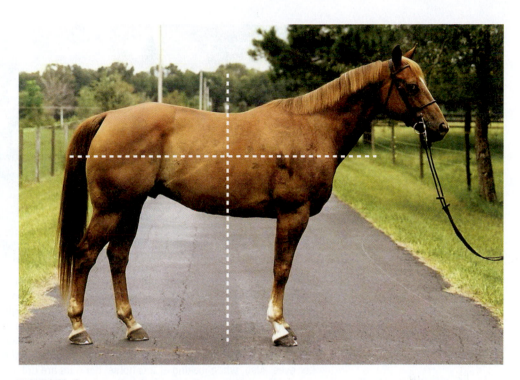

FIGURE 15-1 When evaluating a horse's conformation, it is important to examine the horse from the side and determine if he is carrying his weight evenly from head to tail and from top to bottom. *(Delmar/Cengage Learning. Photo by Kylee Duberstein.)*

Muscling is also an important consideration when evaluating the horse. The quantity and quality of the muscle are evaluated when looking at the horse from sides, front, and back.

Quality of the horse refers to subtleties such as refinement. Refinement in horses means that the horse has a very clean, neat look including chiseled features, well shaped ears, nose and muzzle, and a general lack of coarseness. This does not usually affect the performance of the horse but gives the horse a "prettier" appearance.

THE HEAD AND NECK

When first evaluating a horse, it is important to begin by analyzing the structure, proportions, and angles of the horse. The goal is to have a balanced horse with correct angles and ratios of different body parts to allow for maximum power and maneuverability. Let's begin by examining the horse's head. The ideal head will vary among breeds since different breeds have certain characteristics specific to them. However, a few basic principles pertaining to the size and shape of the head will predominate across all breeds. First, we need to examine ratios that determine the conformation of the head. The distance from the poll to the midpoint between the eyes should be half the distance from the midpoint of the eyes to the midpoint of the nostrils. In other words, the eye will be positioned at about one-third the distance from the poll to the nostrils. Also, the width of the horse's head from the outside of one eye to the outside of the other should be approximately the same length as the distance from the poll to a horizontal line drawn between the eyes, figure 15-2. This width is important to provide room for the brain and also for sinuses, tear ducts, and breathing canals located beneath the skull.

(A) 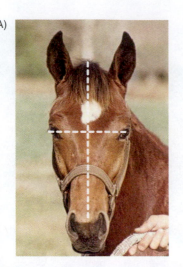 (B)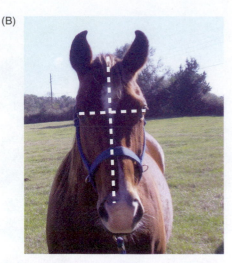

FIGURE 15-2 (A) When looking directly at the front of the horse's head, its eyes should be positioned approximately 1/3 of the way down its head when measured from the poll to the muzzle. The width between its eyes should be the same length as a line drawn from the poll to a horizontal line between the eyes. (B) This horse has a head that is very long in proportion to its width. Her eyes are placed approximately 1/3 of the way down her head but the distance between her eyes is not a great as the distance from her poll to a horizontal line drawn between her eyes. Also, when looking at this horse, her ears are not very neat and refined. Notice the way her ears turn in at the tips; detracting from a refined look. *(Delmar/Cengage Learning. Photo by Kylee Duberstein.)*

The major function of the head, apart from essential purposes such as breathing and eyesight, is to serve as a pendulum to balance the horse as it moves. Therefore, it is critical for the head to be proportionally sized to the rest of the horse's body. The horse's head is heavier in proportion to the length of neck than any other animal in the world. The ideal head of an average sized horse weighs approximately 40 pounds. If the neck is the proper length and the head is proportionally sized to the rest of the body, it can function as a balancing aid for the horse's movement. A classic example of this is the way a horse moves when it injures a leg. When a horse is injured in a foreleg, he limps by using his head to pull his body upward; he lifts his head and pulls in the opposite direction of the injured leg. When he has injured a hind leg, he lowers his head and pulls in opposite direction of the back leg. The head is acting as a pendulum to take weight off of the injured leg. When a horse has too large and heavy a head, it tends to be unable to move weight off of its front end and therefore lacks in athletic ability. A horse with too small a head in proportion to its body tends to be very light on the front end and bouncy; there is too much up and down movement and not enough forward motion.

Other structural characteristics of the head that are generally faulted are the Roman nose and the platter jaw. A Roman nose describes a condition in which the front of the horse's face is rounded outward as opposed to being flat, figure 15-3. This usually does not affect the use of the horse other than it is not as attractive and often adds weight to the horse's head. It is undesirable simply because it takes away from a refined, chiseled look associated with show quality animals. A platter jaw is a condition that describes excessively large jaws on the horse. It also detracts from a refined look and is undesirable because it adds weight and interferes with the horse's ability to flex at the poll.

(A)  (B)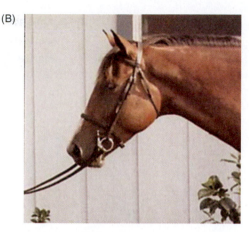

FIGURE 15–3 The horse in Figure 3A has a roman nose. Note that the nose is slightly rounded outward between the eyes and muzzle due to excess bone in this area. The horse is Figure 3B has a very neat and refined head and has a very flat line from eye to muzzle. *(Delmar/Cengage Learning. Photo by Kylee Duberstein.)*

Other important considerations when examining the head are nostril size, and eye size and shape. Nostrils should be large and round to allow maximum intake of air when the horse is working hard and breathing heavily. It is also desirable for the horse to have large, dark eyes set far apart and to the outside of the head to allow for good vision. When looking at the head, you will notice a large depression in the skull above each eye. This is normal and functions to protect the eye from impact. The horse may not be injured even by a severe blow to the eye because the eye and cushion of fat behind the eye is able to recede into that depression, thereby absorbing the impact.

It is important to understand the horse's field of vision in order to understand why eye placement on the head is important. A horse has more developed monocular vision than binocular vision. This means that the horse sees a different picture out of each eye (monocular vision) very well but cannot see as clear a picture with both eyes directly in front of it (binocular vision). Horses have circular vision, and each eye has its own circular field, figure 15-4. These circles barely meet at a point out in front of the horse. The horse cannot see what is directly in front of it. It must turn or lift its head to see objects within 4 to 6 feet of it. It also cannot see directly behind itself. Because of these factors, horses with small eyes, or eyes that are too close together, are faulted due to the fact that their field of vision may be more limited.

When looking at a horse's neck, the ratio of the top of the neck to the bottom of the neck and the ratio of the throatlatch to the head length are critical to consider. The top line of the neck is measured from the poll to the withers and the bottom line is from the throatlatch to the shoulder junction. The ideal ratio would be a 2:1 top line:bottom line ratio. This allows the horse to have a more sloped shoulder since the withers will be set back well behind the point of the shoulder. This also allows the horse to flex at the poll and carry his neck in a slight arch, figure 15-5. A horse that has a longer bottom line than top line is said to be "ewe necked." This is a very undesirable conformation trait as it typically is associated with a straight shoulder and a lack of ability to flex and lower the head. One other ratio that is important to consider is the ratio of the throatlatch to the length of the head. The throatlatch is measured from the poll to the windpipe of the horse and should be roughly half the length of the head as measured from the poll to the muzzle, figure 15-6. If the throatlatch is longer and thicker than this, it restricts the horse from flexing his poll. Horses with deep, coarse throatlatches can possibly

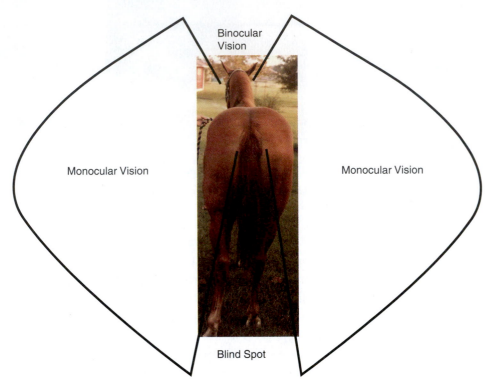

FIGURE 15–4 This diagram shows the vision range of the horse using monocular and binocular vision. The horse also has a blind spot directly behind it and directly under its head. *(Delmar/Cengage Learning. Photo by Kylee Duberstein.)*

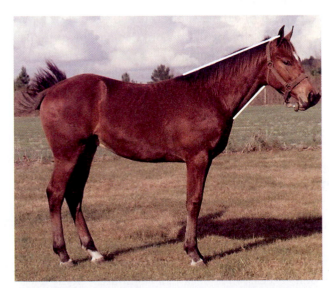

FIGURE 15–5 For a horse to carry its neck properly and flex at the poll easily, the topline of the neck (top white line—distance from poll to withers) should be approximately twice as the underline of the neck (bottom white line—distance from throatlatch to chest). *(Delmar/Cengage Learning. Photo by Kylee Duberstein.)*

FIGURE 15–6 In order for the horse to easily flex its neck at the poll, the distance from its poll to the underside of its throatlatch should be only about half the length of the distance from its poll to its muzzle. *(Delmar/Cengage Learning. Photo by Kylee Duberstein.)*

have trouble breathing when asked to flex their head toward their chest. This also detracts from a refined show quality appearance.

SHOULDER, BACK, AND HIP

After closely examining the horse's head and neck, it is important to begin to look at the proportions of the rest of the horse's body. To begin with, the horse should carry equal weight on his front end and back end and on his top line and bottom line. The neck, shoulder, back, and hip should all be approximately equal lengths, figure 15-7. In addition, it is desirable for the horse to have a deep heart girth. The length of the neck is critical because it directly influences the slope of the horse's shoulder. Slope of the shoulder is measured from the withers to the point of the shoulder. If a line is drawn from the withers to the point of the shoulder and another line is drawn through the withers perpendicular to the ground, as in figure 15-8, the ideal shoulder angle is approximately 45 degrees. The slope of the shoulder directly influences the horse's stride length and smoothness. Too straight of a shoulder causes the horse to have a very short, jarring stride. Horses with a nicely sloped shoulder have a free flowing, smooth, long stride since they are able to reach further with their front legs.

The length of the neck is critical in determining shoulder slope. A short necked horse has withers that tie into the neck much further forward than a longer necked horse. Therefore, the short necked horse often has a very straight shoulder and therefore a short and choppy stride. It is, however, also possible for the horse to have too long a neck. This also creates problems because the long necked horse generally does not have adequate muscling to support the weight of his head and neck. A horse with too short a neck lacks flexibility and maneuverability of the neck, while a horse with too long a neck does not have the strength and coordination to carry the weight of its head and properly use it as a pendulum to balance.

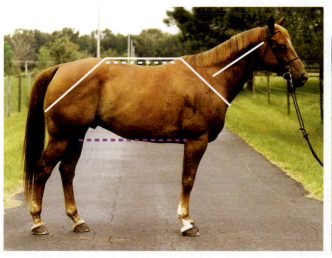
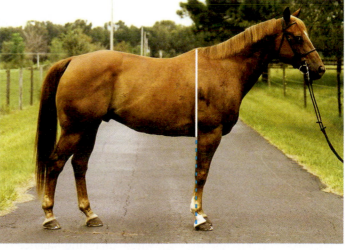

(A)   (B)

FIGURE 15–7 (A) When viewing the horse from the side, the length of the neck, shoulder, back and hip should be approximately the same (all white solid lines). The length of the horse's back (white dashed line) should be approximately one half the length of the underline of the horse (purple dashed line). (B) It is important for the horse to have a deep heart girth to allow maximal room for heart and lung capacity. Heart girth is measured from the withers to the floor of the horse's underside (white line) and should be the same length as the distance from the underside of the horse to the ground (blue dashed line). *(Delmar/Cengage Learning. Photo by Kylee Duberstein.)*

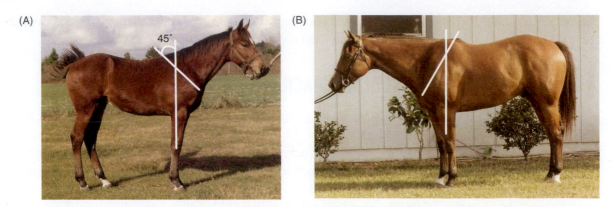

FIGURE 15–8 Horses should have a shoulder that is well sloped or "laid back". The horse in figure 8A has a very good shoulder angle while the horse in figure 8B has a shoulder which is too steep (i.e. the angle of the shoulder is less than 45°). *(Delmar/Cengage Learning. Photo by Kylee Duberstein.)*

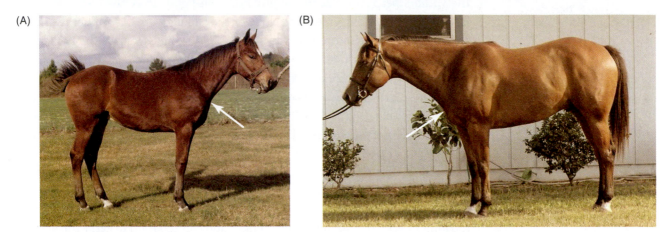

FIGURE 15–9 The horse in figure 9B has a neck that ties in high to its chest and shoulder. This allows this horse to have a greater slope to its shoulder and also a trimmer, more refined neck. The horse in figure 9B, which we have already shown to have too steep a shoulder, has a neck that ties into its shoulder and chest very low. Notice how much thicker and heavier its neck looks compared to the horse in 9A. *(Delmar/Cengage Learning. Photo by Kylee Duberstein.)*

Another important consideration when examining the horse's neck and shoulder is the point where the neck ties into the shoulder. It is preferred that the horse's neck tie in high to its shoulder. This typically will allow for greater slope to the shoulder and a neater, more refined neck. If the horse's neck ties in low, the neck tends to be much heavier and the shoulder is usually straighter, figure 15-9. Thus, it is important to recognize that the neck ratios and position are important in determining the position of other parts of the body and play a critical role in both balance and stride length.

When evaluating a horse's back length, a common flaw seen is the horse with an excessively long back in relation to its neck and hip. An important ratio to consider when analyzing balance is the ratio of the top line to the under line. The top line is measured from the withers to the point of coupling. The under line is measured from a point under the belly between the horse's front legs to a point roughly even with the stifle, figure 15-10. The top line should always be shorter than the bottom line. A longer top line indicates that the horse has a long, weak back. This is problematic because long backs tend to have weaker muscling, and also the longer length makes it difficult for the horse to bring its hind legs up under its body when it moves. The hind legs reaching

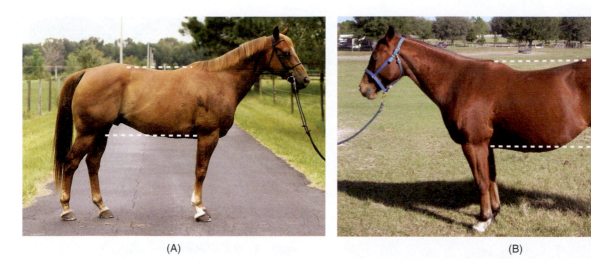

FIGURE 15-10 A short, strong back is essential to allow the horse to easily balance on its hind quarters and bring its hind legs well underneath its body when moving. The horse in figure 10A has a very desirable topline to bottomline ratio of 1:2 whereas the horse in figure 10B has a topline that is almost as long as her underline, making her back weak. The horse in figure 10B will have a much harder time balancing her weight on her hind end as compared to the horse in figure 10A. *(Delmar/Cengage Learning. Photo by Kylee Duberstein.)*

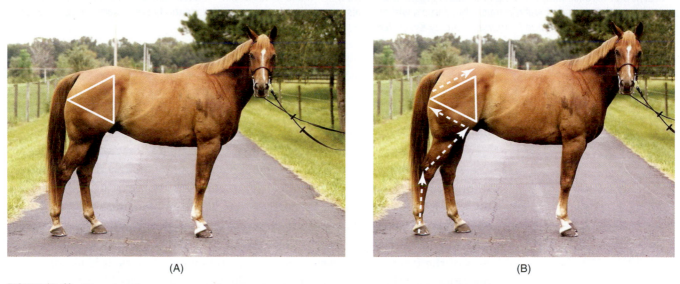

FIGURE 15-11 When looking at a horse's hip from the side, you should be able to draw an equilateral triangle from the point of the hip to the point of the rump to the stifle (A). This allows for maximal power as the horse moves. When the horse's hind foot pushes off the ground, force is transmitted up to the hock, then to the stifle, next to the point of the rump, and finally across the topline (B). If any of the sides are disproportional, power is lost as the force travels up the leg. *(Delmar/Cengage Learning. Photo by Kylee Duberstein.)*

under the horse's body are the source of power for the horse to move forward and allow the horse to maneuver and adjust easily. If a horse is unable to bring its hind legs well underneath its body, more weight must be carried on its front end, thereby reducing its power and maneuverability.

The length of the hip is also critical to a horse's power. In general, larger hips are better. A hip should be approximately the same length as the back. The hip and hindquarters are the driving power behind the horse. Almost all disciplines of riding have maneuvers that require power and adjustability. The larger and better the hip is, the more power the horse will have. It is also important to consider the way the hip is shaped. On the ideal hip, one should be able to draw an equilateral triangle from the point of the hip to the muscling just above the cheekbone of the rump to the point of the stifle joint, figure 15-11. This allows for maximum force generation. As the horse moves, it is not a simple forward movement. The horse

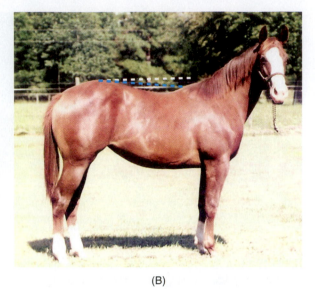

(A)   (B)

FIGURE 15–12   Figure 12A shows a horse that is properly balanced at the withers and hip. A straight line (parallel to the ground) can be drawn that touches both the hip and withers. Figure 12B shows a horse that is "downhill". If a line is drawn from the hip that is parallel from the ground, it will not touch this horse's withers. This is undesirable because it forces the horse to carry more weight on its front end. *(Delmar/Cengage Learning. Photo by Kylee Duberstein.)*

actually pushes down against the ground and that force goes through the hock and up to the stifle. Because of the attachment of the muscle, the force is then transmitted to the rump cheek and over the top line of the hindquarter. If any one of the sides of the triangle is shorter, there is a loss of force. Correct angles and proportions allow for maximum force generation.

Another important consideration when determining a horse's balance is the hip and wither height. Hip and wither height should be approximately the same. Often, a horse can be slightly uphill, or higher at the withers, without being penalized. If a horse is downhill, or lower at the withers than at the hip, the horse will carry more weight on its front end and lack maneuverability and driving power from behind, figure 15-12. Carrying too much weight on the forehand can also lead to future lameness of the front legs. However, when evaluating young horses, it is important to remember that they will grow faster at the hip than at the withers. Therefore, a young horse may be downhill as it is growing but may catch up as it reaches maturity.

THE LEGS

Another very important consideration when evaluating a horse's structural correctness is the structure and position of the legs. This is critical because the horse's legs take incredible impact in most riding disciplines. Any conformational flaw causes deviations in where the horse absorbs concussion. Conformational defects affect the horse's way of moving and also lead to future lamenesses due to excessive stress placed on certain areas of the body. A horse carries approximately 65% of its weight on its front legs. Therefore, front leg injuries resulting from trauma or concussion are the most prevalent type of leg injury. Conformation impacts this because conformational defects cause deviations in the way the horse moves and places its hooves on the ground, and therefore the way impact travels up the leg. The more structurally correct the horse's legs are, the more evenly distributed the impact will be and the less likely the horse will be to have injuries and lamenesses.

Blemishes and Lamenesses

Before we can further discuss conformational defects, we must first go into detail on some common injuries and blemishes found on horses. **Blemishes** are conditions that cause a noticeable change on an area of the horse but most likely do not impact the way a horse can perform. Figure 15-13 will illustrate common blemishes and injuries found on the horse's legs. A common blemish found on the lower legs of the horse is a **splint**. This is a hardened lump formed in the area of the splint bone (a non-weight-bearing bone found in the lower leg) due to either injury to the splint bone or to the ligament that attaches the splint bone to the cannon bone. This condition can either be caused by a trauma to the region or by repeated concussion. Usually this does not cause lameness once inflammation is reduced and is considered a blemish that does not hinder performance.

Another such blemish is a **windpuff** or windgall. These are soft swellings slightly above the fetlock which are caused by excess synovial secretion from the fetlock joint. These can be caused by jarring of the leg which causes excessive synovial fluid to be secreted to lubricate the area where the tendon passes around the sesamoid bone. It typically does not cause lameness but is an unsightly blemish. Injuries that can be observed on the horse that do impair performance include **ringbone**, **sidebone**, and **bowed tendons**. Ringbone and sidebone both involve the formation of excessive bone in the joint region. Ringbone is a bony enlargement of the bones or joints in the pastern. This can cause lameness and pain in the area. Sidebone is bone formation on the cartilage of the side of the lower pastern joint. It is typically caused by injury or concussion causing calcium to accumulate and harden along the joint. Ringbone, sidebone, and splints can all occur on front or hind legs.

A bowed tendon is also a visible condition that causes lameness and impaired performance. This typically occurs on the front legs but can occur in hind legs as well. It appears as a bowlike swelling on the back of the leg between the knee and

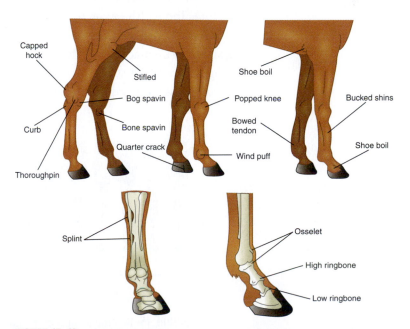

FIGURE 15–13 Diagram of location and appearance of potential unsoundnesses and blemishes found in the horse's legs. *(Delmar/Cengage Learning.)*

the ankle caused by tearing of the superficial digital flexor tendon. It heals very slowly and the tendon is replaced by scar tissue.

There are also a few conditions that are unique to the hind legs. **Capped hock** refers to bursitis in the hock. It is an accumulation of fluid on the back of the hock that is typically caused by injury. It usually does not cause lameness but will be a permanent blemish unless treated early and aggressively. A **curb** on a hock can look somewhat similar. It is a firm swelling on the hock approximately 4 inches below the point of the hock. It is caused by strain on the ligament connecting the hind cannon to the hock. It does not generally cause permanent lameness in the horse. A **bog spavin** is another condition of the hock. It is a soft bulge on the outside of the hock caused by the accumulation of excess synovial (joint) fluid in that area. This is caused by inflammation of the membrane covering the joint but does not generally cause lameness. A bone spavin is located on the inside of the hock and is a bony bump caused by inflammation to the periosteum. This causes excess bone to be deposited over the inside area of the hock joint. The damage and inflammation associated with this condition are the result of compression and rotation stresses on the hock.

All of these blemishes and lamenesses can be caused either by injury or repeated aggravation to the area. This can be due to excessive training or also to poor conformation that puts increased stress on a particular area. When examining a horse with blemishes, it is important to analyze the horse's structure to determine if poor conformation potentially played a role in the formation of the blemish or injury.

Front Leg Conformation

When analyzing the conformation of the horse's legs, it is important to look at the legs from both a front or hind view and a side view. When observing the front legs from a front view (facing the horse), one should be able to draw a straight line from the point of the shoulder to the ground that bisects the leg exactly in half, figure 15-14. The hoof and knee should point forward and also be bisected in half by the line. The width of the hooves on the ground should be exactly the same as the width of the legs as they originate from the chest. Deviations from this cause extra strain to be placed on different regions of the legs.

Figure 15-14 illustrates common leg conformational defects. The first horse shows correct leg conformation from the front. Notice that the line bisects the entire

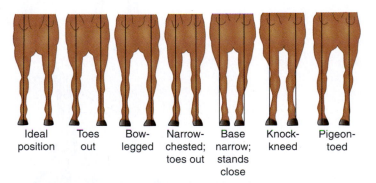

FIGURE 15–14 Frontal view of the horse's front legs. This diagram shows both correct conformation and common deviations observed in the horse. (*Delmar/Cengage Learning.*)

leg into two equal portions and also that the knee and foot point directly forward. In the next sketch, you see a horse that is splayfooted or toes out. The knees are set to the inside of the line and the toes point outward. The next horse is bow-legged. This horse stands with its knees set outside of the straight line. In the next picture, the horse has a narrow chest and toes out. The next horse is base narrow; it stands with its hooves very close together in relation to where its legs leave its chest.

Horses can also be base wide to where the horse's feet are much farther apart than the legs are at the point where they come out of the chest. Knock-kneed horses have knees that are set to the inside of the straight line. A pigeon-toed horse has straight, correct legs until you notice that its toes point inward severely. This is caused by a rotational deviation of either the cannon bone, fetlock joint, or pastern bones.

It is important to understand that all of these conformational defects can lead to lamenesses and blemishes, because they cause excessive pressure to be centered on a certain area of the legs. For example, base narrow horses are predisposed to landing on the outside of their hoof wall. This pressure on the outside of the hoof wall can lead to conditions such as ringbone, side bone, and heel bruising. At the opposite extreme, base wide horses tend to also toe out. This causes weight to be distributed more on the inside of the horse's hoof and also predisposes them to ringbone and sidebone.

Deviations of the knee cause increased strain on this joint and the ligaments and tendons attached to it. Bowlegs (knee set to the outside) cause increased tension on the outside of the leg. Knock knees (knees set to the inside) tend to also be associated with pigeon toes and therefore have rotation of either the cannon bone, fetlock, or pastern. This causes extreme pressure to be placed on the point of rotation. All of these deviations cause an unequal line of concussion. As the horse moves and its foot strikes the ground, the impact travels up the leg in an uneven line. The concussion from the impact is not evenly distributed. The area that absorbs more of the concussion is more likely to be damaged.

When examining the horse's front legs, you should also look at the legs facing the side of the horse, figure 15-15. A straight line should be able to be drawn from the center of the scapula through the front edge of the knee that bisects the hoof. Structural deviations that may be observed are camping out and camping under.

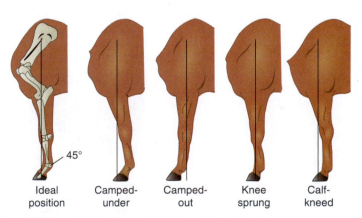

FIGURE 15–15 Side view of the horse's front legs, again showing correct conformation and commonly seen deviations. *(Delmar/Cengage Learning.)*

A horse that is camped out will stand with its legs too far in front of it. This causes excessive pressure to be placed on the knee and fetlock joint as these joints must almost bend backwards for the horse to stand in this position. It also places excessive stress on the hooves. This stance can be due to conformational defects or sometimes due to hoof pain, such as navicular. The horse that is camped under will stand with its legs too far underneath it. This causes increased strain to be placed on the ligaments and tendons of the leg. It also causes the horse to carry its weight too far forward (too much weight on its front end) which can cause lameness due to stress and also causes the horse to have a short, choppy stride and potential stumbling.

Two other conditions that may be observed from the side of the horse are calf knees and buck knees (knee sprung). If the line does not bisect the knee but instead is to the front of the knee, the horse is considered calf kneed. This places excessive strain on the back ligaments and tendons of the leg, as well as pressure on the joints. If the line is in back of the knee, the horse is considered over at the knees or buck kneed, a condition that also distributes pressure unequally over the leg.

Hind Leg Conformation

Similar lines and angles can be drawn to evaluate the hind legs of the horse. You should be able to draw a straight line from the horse's rump cheek, through its hock, and through its fetlock when facing the hind quarters from behind the horse, figure 15-16. The hooves on the back leg, however, point slightly outward. The first horse in figure 15-16 shows a horse with correct leg structure when viewed from behind. A horse can also stand base wide (second horse) or base narrow (third horse) on its hind legs. The hocks can also deviate from being straight. A cow hocked horse has hocks that point inward. Problems associated with horses that are either cow hocked or base narrow include additional stress placed on the leg and joints (bog and bone spavins can be associated with cow hocks), as well as potential interference when the horse moves. In other words, since the entire leg or part of the leg is closer together than normal, the horse is more prone to hitting its legs against each other as it moves. Horses that are bow legged have the opposite condition of cow hocked horses. These horses have hocks that turn outward when viewed from behind. Bow legged and base wide horses often have trouble being able to properly use and push off of their hind legs and therefore can lack the athletic ability of a horse with proper conformation.

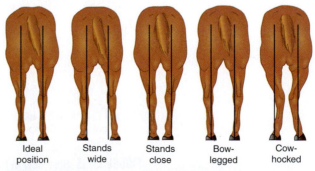

| Ideal position | Stands wide | Stands close | Bow-legged | Cow-hocked |

FIGURE 15–16 Hind view of hind legs with correct conformation and commonly seen deviations. *(Delmar/Cengage Learning.)*

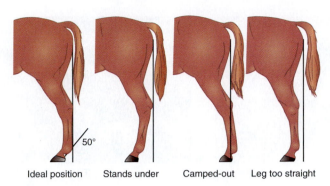

Ideal position   Stands under   Camped-out   Leg too straight

FIGURE 15–17 Side view of hind legs showing correct conformation and commonly observed deviations. *(Delmar/Cengage Learning.)*

When examining a horse's hind legs from a side view, you should be able to draw a line perpendicular to the ground that touches the point of the horse's rump cheek, the back of the hock, and the back of the fetlock, figure 15-17. This conformation of the hind leg allows the horse to properly carry weight on its hindquarters and reach under itself with its hindquarters as it moves to allow for maximum power. A horse that is sickle hocked has too much angle, or set, to its hocks and tends to stand under (second horse in figure 15-17). This puts extreme stress on the joint and surrounding tendons and ligaments and can lead to conditions such as curbed hocks, bog spavin, and bone spavin. Horses can also stand camped out, where their hind legs are naturally behind them. This condition makes it very difficult for the horse to use its hindquarters well. Horses that have the opposite conformation of a sickle hocked horse are said to be post legged (such as the fourth horse in figure 15-17). These horses have extremely straight angles to their hocks. This puts extreme strain on the hock and can also cause bog spavins and bone spavins.

Lower Leg Conformation

A final important consideration when examining the horse's legs is the angle and length of the horse's pastern. The pastern acts as a shock absorber for impact from the hoof landing on the ground. Therefore, the conformation of the pastern will affect the soundness of the ankles, knees, and hocks. The pastern angle typically matches the shoulder angle and should be approximately 45 degrees when measured from the ground, figure 15-18. The pasterns must be long enough and sloped enough to absorb impact, but if they are too long or too sloped, the fetlock will hit the ground upon severe impact, as shown in the figure. This can cause fractures in the sesamoid bones at the back of the fetlock. However, too straight a pastern angle (often resulting from pasterns also being too short) causes increased jarring of the leg and joints when the horse moves and therefore can lead to windpuffs, enlargement of the sesamoid, and joint pain, as shown in the figure. Pasterns that are too straight also affect the navicular bone. Too straight an angle causes the navicular to come in contact with the short pastern bone leading to erosion of the bone or the formation of bony spurs. Finally, the pastern angle and the hoof angle should be the same. A horse with a very steep hoof angle when compared to its pastern angle is said to be club footed, figure 15-19. This is undesirable because

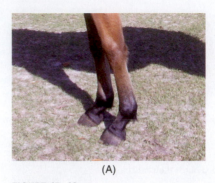
(A)

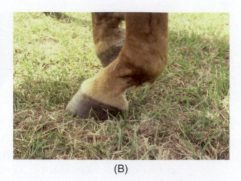
(B)

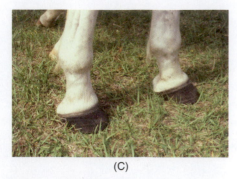
(C)

FIGURE 15–18 Pastern angles. Figure 18A shows a correctly sloped pastern, while Figure 18B shows a pastern with too much slope and Figure 18C shows a horse with pasterns that are too straight. *(Delmar/Cengage Learning. Photo by Kylee Duberstein.)*

FIGURE 15–19 Club foot. The front foot is clubbed; note the steep angle of the hoof. *(Delmar/Cengage Learning. Photo by Kylee Duberstein.)*

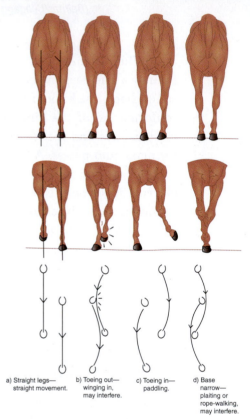

a) Straight legs—straight movement.

b) Toeing out—winging in, may interfere.

c) Toeing in—paddling.

d) Base narrow—plaiting or rope-walking, may interfere.

FIGURE 15–20 Diagram showing how front leg conformation affects movement in the horse. *(Delmar/Cengage Learning.)*

the steep angle of the horse's hoof will not only change the way it moves but also make the horse prone to foot and leg lameness on the affected leg.

Movement

It is important to recognize that leg conformation significantly impacts the way the horse moves. A horse with straight, correct legs can reach easily with its legs as it moves and also places them on the ground cleanly and correctly without any interference. Figure 15-20 shows front leg conformation and how it affects movement. Notice that if the legs are straight, then as the horse moves the legs are placed straight on the ground and do not interfere (in other words, they do not hit against each other at any point in time).

FIGURE 15–21 Extension of the front leg through the shoulder. Note how the horse in this picture has very little flexion at the knee. *(Delmar/Cengage Learning. Photo by Kylee Duberstein.)*

Horses with structural deviations in their legs do not generally move their legs straight forward when traveling. Horses with pigeon toes typically wing out when they move. As the horse has its knee bent and its leg brought back behind it, it must swing its leg to the outside of a straight line to place it back in front of itself. This is due to the natural angle that the horse's legs are set at due to the pigeon toes. It is not, however, as serious a problem as the horse that toes out. These horses will wing in as they move forward. This causes the horse to potentially interfere and hit its other leg as it is moving one leg forward. Horses that are base narrow tend to cross one front leg over the other when moving, thereby also having a tendency to interfere. When analyzing the horse's leg conformation, recognize that the correctness of the leg greatly influences the way the horse will place its legs down as it moves.

In addition to watching the horse from the front and rear to determine its footfall, it is also important to watch the horse move from the side. This enables the judge to look at the horse's stride length. In most disciplines and breeds, you want the horse to have a long, smooth stride that is very flat with very little knee action, figure 15-21. For certain breeds such as Arabians, Morgans, and Saddlebreds, you want the horse to have more knee flexion and raise its legs higher, figure 15-22. It is important for all horses to bring their hind legs well underneath themselves to power their movement. It is also important when watching the horse move from all angles to be sure that the horse does not interfere or hit its legs together at any point in its stride.

MUSCLING IN THE HORSE

The quantity and quality of muscling is also an important criterion when evaluating a halter horse. To evaluate this, a judge must examine the horse from all sides. While looking at the front of the horse, the judge will analyze the amount of

FIGURE 15–22 A horse with much more flexion at the knee (and the hock) than the horse in the previous figure. *(Delmar/Cengage Learning. Photo by Kylee Duberstein.)*

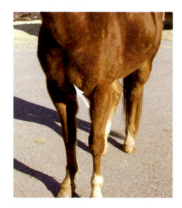

FIGURE 15–23 The arrow indicates the pectoral V of the horse's chest. It is desirable for this to be well defined with substantial muscling on each side of the V. *(Delmar/Cengage Learning. Photo by Kylee Duberstein.)*

muscling in the chest and forearm. It is desirable for the horse to be heavily muscled versus average or light in those areas. It is important to recognize that some breeds are more heavily muscled (e.g., American Quarter Horse) than others (e.g., Thoroughbred). Quality of muscling is also important, and the muscle should be well defined rather than flat. A deep, pectoral V shape is desirable, figure 15-23. It is also desirable for the forearm and gaskin muscling to be long and smooth versus short and bunchy. When examining the horse from the side, the muscling over the back and loin area should be smooth and defined rather than weak. The back should tie smoothly into the hip without severe angles or bumps. The muscling over the entire top line should be smooth and flow together seamlessly, figure 15-24.

On the hindquarters, the muscling over the stifle and gaskin should also be long rather than short. The muscling around the stifle should be wide. In fact, the stifle muscling should be the widest part of the horse when viewed from behind, figure 15-25. The muscling around the inner and outer gaskin should also be wide and well defined. In general, it is desirable to have a smooth, well-defined muscle pattern over the entire horse. Quantity of muscle will at least in part be determined by breed.

QUALITY OF THE HORSE

Quality in a horse refers to refinement, smoothness, and elegance or style. Refinement is a general lack of coarseness or excess tissue/bone in a horse. A refined horse has very chiseled, trim features with clean bone structure (no protrusions, bumps, etc). Smoothness refers to the hair coat and muscle pattern. The hair coat of a show horse should be short and lie flat, and the muscle pattern should be smooth. All parts of the horse's body should flow seamlessly together with no rough edges or protrusions where the parts of the body meet. The top horse should possess a certain style which is seen through its mannerism and self-carriage. The horse should look attractive, pleasant, and attentive while standing and moving.

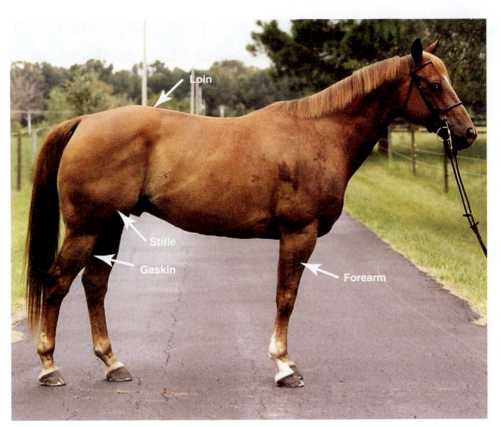

FIGURE 15–24 The arrows indicate areas where muscling is typically evaluated. *(Delmar/Cengage Learning. Photo by Kylee Duberstein.)*

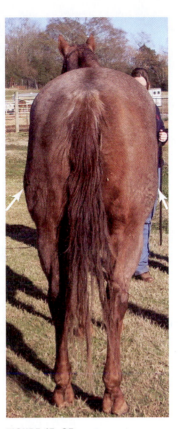

FIGURE 15–25 The arrows indicate the muscling that will cover the stifle area. When viewing the horse from behind, this should be the widest area on the horse. *(Delmar/ Cengage Learning. Photo by Kylee Duberstein.)*

This is sometimes referred to as presence, and describes how the horse should carry itself when being judged. Common characteristics that detract from refinement are Roman noses, platter jaws, excessively coarse bone structure in the legs, and protrusions where the back ties into either the withers or the hip.

SUMMARY

Halter horse judging involves analyzing a particular breed and type of horse for balance, structural correctness, muscling, type, way of moving, and quality. Proper conformation is important to allow the horse to be balanced, powerful, and maneuverable. Correct conformation is also critical to keep the horse sound over its lifespan. Conformational flaws in the legs result in uneven distribution of pressure. Oftentimes, the area that absorbs the majority of the impact is the area that is damaged. Proper ratios and angles of the horse's body parts are also important to allow the horse to carry itself in a balanced manner and to be able to perform with maximum power and athletic ability. Evaluating a horse based on its conformation should give you an idea of how the horse might perform a given task and how sound it will stay over its productive lifespan. When judging a horse of a particular breed, you should consider both the horse's conformation and how well the horse represents the ideal horse of that particular breed.

STUDENT LEARNING ACTIVITIES

1. Look at a real horse or realistic picture of a horse. Evaluate the horse's conformation based on the criteria presented in this chapter.
2. Look at four real horses or realistic pictures of horses. Rank the horses from first to fourth and write a set of oral reasons to defend your placing. Present these to the class.
3. Take a diagram of a horse and label common blemishes found on the horse.

 ## TRUE OR FALSE

T   F   1. Hunter horses tend to be taller with a longer, leaner muscle pattern and a longer stride as compared to western horses.

T   F   2. Balance refers to equal distribution of muscling and weight from the front of the horse to the back of the horse, from its top to its bottom, and from side to side.

T   F   3. Structural correctness refers to how much bone density a horse has in its legs.

T   F   4. The term *type* in horses refers to how well that horse represents its particular breed.

T   F   5. When examining a horse's muscling, quantity is more important than quality.

T   F   6. A horse carries approximately 65% of its weight on its front legs.

T   F   7. Horses with pigeon toes typically wing in when they move.

T   F   8. Arabians, Morgans, and Saddlebreds tend to have more knee action when compared to Quarter Horses, Thoroughbreds, and Warmblood breeds.

T   F   9. A pectoral V shape can be viewed when looking at the horse's shoulder from the side.

T   F   10. Roman noses and platter jaws are marks of refinement in the horse.

 ## FILL IN THE BLANKS

1. Typically, there are two main groups of halter classes: _____ and _____.
2. The major function of the head, apart from essential purposes such as breathing and eyesight, is to serve as a _____ to balance the horse as it moves.
3. A _____ describes a condition in which the front of the horse's face is rounded outward as opposed to being flat.
4. The horse has two types of vision: _____ and _____.
5. A horse that has a longer bottom line than top line is said to be _____.
6. The slope of the horse's shoulder should be approximately _____ degrees.
7. On the ideal hip, one should be able to draw an _____ from the point of the hip to the muscling just above the cheekbone of the rump to the point of the stifle joint.
8. _____ are conditions that cause a noticeable change on an area of the horse but most likely do not impact the way a horse can perform.

9. A _____ is a soft swelling slightly above the fetlock which is caused by excess synovial secretion from the fetlock joint.

10. The horse's pastern acts as a _____ for impact from the hoof landing on the ground.

DISCUSSION

1. Describe the correct conformation of a horse's head and neck.
2. How does the horse use its head for balance when it is lame?
3. If a horse is balanced, what does this mean?
4. Describe why a horse's shoulder angle is critical for movement and soundness.
5. Why is it not desirable for a horse to have a long back in comparison to the rest of its body?
6. Why do structural deviations in the horse's legs predispose a horse to lameness or blemishes?
7. How does pastern angle affect soundness?
8. How does a horse's leg conformation affect its movement?
9. What do we look for when examining a horse's muscling?
10. Describe refinement as it pertains to horse conformation.

CHAPTER 16
Riding Classes

OBJECTIVES

As a result of studying this chapter, students should be able to:

- Name and describe the different gaits most horses possess
- Name and describe gaits that only certain breeds exhibit
- Describe how different disciplines of riding originated
- Describe what conformational traits might make a horse more desirable for a performance
- Describe the criteria for which a hunter under saddle and hunter over fences horse is judged
- Describe how a western pleasure horse should perform
- Explain what horsemanship and equitation are and how they are judged
- Name and describe different western performance events

PERFORMANCE CLASSES

Halter/conformation classes are popular competitive events in the horse industry and are an integral part of many horse shows. However, riding or performance classes are also a common way to compete at horse shows. There are many different types of performance classes available to compete in. Most can be divided into two main disciplines of riding: English events and western events. English disciplines originated from foxhunts in England. The riding is traditionally more of a forward seat where the rider's body is still centered over the middle of the horse but is inclined forward to move with the horse. These horses traditionally covered lots of terrain and needed to have big, smooth strides with the ability to jump obstacles safely. Today, there are many different English events, but they all emphasize good movement and athletic ability.

Western disciplines originated from the days when horses were used to work cattle. These horses had to be comfortable to ride all day long and had to be extremely obedient in order to be useful in handling cattle. Today, there are many different western events ranging from classes that judge how well a horse moves and obeys to classes that judge the horse's ability to work cattle in different ways. In horse judging competitions, students will be asked to judge a variety of performance events in both western and English disciplines. Some classes require that the student both places the class correctly and presents an oral reasons defense; other classes only require that the student give the performance a score.

HORSE GAITS

Before learning the different types of classes that might be seen at a horse show or judging competition, one must first understand the basic **gaits** that horses may be required to perform. Typically when judging stock horse breeds (Quarter Horses, Paints, Appaloosas, etc.), and sport horse breeds (Thoroughbreds and Warmbloods), the four gaits seen are the walk, trot, canter, and hand gallop. A **walk** is a slow, four-beat gait in which each hoof falls independently of the other legs, figure 16-1. A **trot** is a medium paced gait (approximately 4 m/s) that consists of two beats; the legs move in diagonal pairs, figure 16-2. The **canter** is another medium paced gait that is roughly twice as fast as the trot (approximately 8 m/s). It is a three-beat gait and generally the horse is required to be on the correct **lead**, meaning that the inside front leg is always the leading leg. Therefore, if the horse is traveling on a circle to the right, the left hind leg moves forward and

FIGURE 16–1 The walk is a natural four-beat gait where each of the horse's legs hits the ground independently of the other legs. *(Delmar/Cengage Learning. Photo by Kylee Duberstein.)*

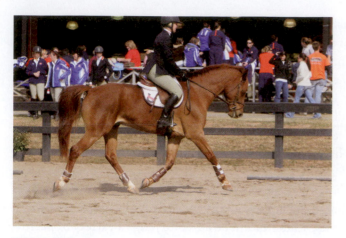

FIGURE 16–2 The trot is a two-beat gait where the horse moves its legs in diagonal pairs. Notice how the left hind leg and right front leg move forward and backwards as a pair as do the right hind leg and left front leg. *(Courtesy of UGA NCAA Equestrian Team.)*

FIGURE 16–3 This horse is traveling on the right lead. The left hind leg will strike the ground independently of the other legs. The right hind leg and left front leg will strike the ground next. Finally, the right hind leg will strike the ground last and will be the leading leg, or the leg that reaches in front of the horse the farthest. *(Courtesy of UGA NCAA Equestrian Team.)*

strikes the ground first, followed by the diagonal pair of the right hind and left front moving forward and striking the ground at the same time, finished by the right front leg moving forward independently and striking the ground last, figure 16-3.

When the horse changes directions, it is required to do some type of **lead change**. A simple lead change involves the horse slowing down to a trot and then picking up the new lead. A flying lead change involves the horse lifting or suspending his whole body into the air and changing his leg position to the other lead.

A **hand gallop** is a four-beat gait that is faster than the canter and requires the horse to lengthen its stride more. It has a similar rhythm and foot fall pattern as the canter, but the horse is moving faster and, therefore, instead of having a middle diagonal leg pair hitting the ground at the same time, each foot strikes the ground independently. For instance, when traveling left, the horse will move forward and strike the ground first with its right hind, then its left hind, then its right front, then its left front, figure 16-4. Other tasks the horse may have to perform are the halt and back. A halt requires that the horse stop completely. To do this correctly, the horse must shift its weight to its hindquarters rather than jamming its front legs into the ground to stop. The back is a movement where the horse moves backwards. It also is a two-beat gait with the horse's legs moving backwards in diagonal pairs (same leg pairs as a trot but backwards). The horse should back willingly and obediently and should easily flex at the poll. In most all disciplines, horses are penalized for excessive head movement, reluctance to move backwards, and not backing in a straight line.

HUNTER CLASSES
Hunter under Saddle

Hunter under saddle classes are judged on the flat (not over jumps) at the walk, trot, and canter. Horses are judged on their way of moving and obedience. These horses originated from the days of foxhunting and should be suitable for riding

FIGURE 16–4 The horse in a gallop. This is a similar gait to a canter except that the middle beat is broken into two beats as each of the horse's legs lands independently.

over long stretches of terrain and athletic enough to be able to jump obstacles (though no obstacles are present in hunter under saddle classes). The horse's **frame** is something that we will discuss when judging riding horses, and this refers to the way the horse carries itself (head carriage, stride length, etc.). The hunter horse should have a long, relatively low frame. The horses should have a long, ground covering stride and very little knee action. The knees should not bend very high but the horse should move forward freely from the shoulder to maximize the amount of ground covered. The poll of its head should be slightly above the height of its withers and the poll should be flexed to where a straight line can be drawn from the poll to the nose that is perpendicular with the ground, figure 16-5.

Horses that are overflexed, figure 16-6, or not on the bit, figure 16-7, are penalized. The horse should be alert but relaxed and should always be obedient and responsive to its rider. It should easily and willingly extend its stride when asked; it may be asked to perform an extended trot or a hand gallop. It should move correctly and freely and appear athletic. A hunter may also be asked to stop and stand quietly and to back (walk backwards) easily. When performing these maneuvers, the horse should be obedient, attentive, and relaxed. The hunter's stride should be smooth and rhythmical, and the horse should not be excessively fast or slow.

Hunter over Fences

In the **hunter over fences** class, the horse is asked to jump a course of obstacles (approximately eight jumps) ranging anywhere from a few inches to 4 feet high, depending on the class and show specifications. The horse is judged on its jumping style, obedience, and way of going. The horse should carry itself in the same frame and have the same desired movement as hunter under saddle. In addition, it should jump in a safe and pleasant manner. It is important that the horse maintains an even and cadenced stride both approaching the fence and upon landing after the fence.

FIGURE 16–5 The hunter horse in this picture is properly flexed at the poll; if a straight line was drawn from poll to muzzle, it would be perpendicular to the ground. Head carriage of the hunter horse will vary due to breed and class. This horse's head would be slightly high for a Quarter Horse hunter. Breeds such as Arabians may carry their neck in even more of an arch than the horse in this picture. *(Courtesy of UGA NCAA Equestrian Team.)*

FIGURE 16–6 This horse is overflexed or overbridled. If a line is drawn from the horse's poll that is perpendicular to the ground, the horse's muzzle would be behind the line. *(Courtesy of UGA NCAA Equestrian Team.)*

FIGURE 16–7 This horse is not on the bit; it is resisting rein pressure. If a line is drawn from the horse's poll that is perpendicular to the ground, the horse's muzzle would be in front of the line. *(Courtesy of UGA NCAA Equestrian Team.)*

FIGURE 16–8 This horse is jumping in good form. It is reaching with its head and neck and it has drawn its knees up well; if a line was drawn from its elbow to its knee, the line would be parallel to the ground. On a side note, the bridle used on this horse is not typical for hunters. *(Courtesy of UGA NCAA Equestrian Team.)*

The horse should form an arc over the middle of the fence and reach with its head and neck rather than stiffening it, figure 16-8. The horse's back should be rounded in the air rather than stiff and flat, figure 16-9. It should lift its knees from the shoulder to clear the obstacle. If the horse is jumping correctly, a line should be able to be drawn from the top of the leg (where the front leg meets the shoulder) to the knee that is parallel to the ground, figure 16-10. The knees should also be even rather than the horse splitting his knees, figure 16-11. The horse is penalized if he jumps with a form that is unsafe such as hanging legs or splitting his knees.

These faults make it more likely that the horse will fall if it hits the obstacle since its legs are more likely to be pulled underneath its body or a pole trapped in its legs. The horse is also penalized if it jumps from too close or too far away from the jump.

FIGURE 16-9 This horse is jumping flat. Notice how stiff it looks through its back and neck. *(Courtesy of UGA NCAA Equestrian Team.)*

FIGURE 16-10 (A) This horse jumps with very correct form. Its knees are even and a straight line can be drawn from the top of its leg through its knee that is parallel to the ground. (B) This horse does not have as correct a form. Notice that a line drawn from the top of its leg to its knee would not be parallel to the ground. *(Courtesy of UGA NCAA Equestrian Team.)*

FIGURE 16-11 This horse is splitting its knees; the knees are not level and one leg is lifted much higher than the other. *(Courtesy of UGA NCAA Equestrian Team.)*

FIGURE 16–12 (A) This horse is leaving from a good take-off point, and (B) is landing in a position to allow the horse to continue moving freely. (C) This horse is taking off much too close to the jump, causing it to either hit the jump with its front leg or stop its momentum to clear the jump and thereby lose its momentum upon landing. *(Courtesy of UGA NCAA Equestrian Team.)*

The ideal take-off point is a distance away from the jump that is roughly equal to the height of the jump (e.g., if the jump is 3 feet high, the take of point would be approximately 3 feet in front of the jump). This allows for the horse to jump comfortably across the jump without losing or gaining momentum, figure 16-12. If the horse leaves from too far away from the jump, it has to push harder off the ground and therefore has an increased momentum as it is landing. In addition, this is unsafe because if the horse leaves excessively far away from the jump, it may not be able to get completely over the jump and may fall into it. If the horse leaves from too close to the jump, it must halt its momentum and push upwards very hard to keep from hitting the jump. This causes it to land with a decreased momentum. It also makes it more likely that the horse will hit the jump with its front legs as it is trying to jump because there might not be enough room to raise its legs without hitting the jump. Overall, the hunter over fences should look smooth and rhythmical with very little change in speed throughout the course. It should jump easily and in good form. It should have a pleasant expression and be very obedient; it should not look as though the rider has to work excessively hard to either slow down or speed up the horse.

Hunter Hack

The hunter hack class is somewhat of a combination of the hunter under saddle and hunter over fences classes. In the hunter hack class, the horse is required to jump two fences, generally between 2 feet and 2.5 feet high. They generally are asked to hand gallop after the jump and then halt. They are then required to walk, trot, and canter in both directions on the rail. The jumps are given more emphasis (70%) than the rail work (30%), and therefore the rail work is mostly used to break ties between horses. The rail work is judged in the same way as a hunter under saddle class would be judged and the jumping portion is judged on the same criteria as the hunter over fences class is judged. The horse should show a definite lengthening of stride and increase in speed during the hand gallop after the jumps. The horse should stop easily and obediently when asked without excessive pulling or head movement.

Huntseat Equitation

Equitation classes are judged on the position and effectiveness of the rider. These can be judged on the flat (under saddle classes) or over fences. Proper body position of the rider requires that the person sit in the middle of the saddle with weight being evenly distributed on each side of the horse. A straight line should be able to be drawn from the rider's ear to hip to heel, figure 16-13. Riders may incline slightly forward when posting a trot or riding a hand gallop. A straight line should also be able to be drawn from the rider's elbow to hand to the horse's bit, figure 16-14. The hands and arms should be relaxed with the thumbs pointing

FIGURE 16–13 (A) When riding correctly, a straight line should be able to be drawn from the rider's ear to shoulder to hip to heel. (B) A rider inclined slightly forward is an acceptable position for riding a posting trot or a hand gallop. *(Courtesy of UGA NCAA Equestrian Team.)*

FIGURE 16–14 This rider has correct placement of her hand and arm; a straight line can be drawn from her elbow to her hand to the horse's mouth. *(Courtesy of UGA NCAA Equestrian Team.)*

inwards at a 30 degree angle. Riders should not move excessively when riding and riders' cues should be virtually invisible to the onlooker. When judged on the flat, the riders are generally asked to execute some type of specified pattern and then proceed to rail work. The rail work is used to break ties after judging the pattern. The American Quarter Horse Association publishes the following judging guide (out of 20 possible points) in their official handbook:

20: Excellent equitation including body position and use of aids. Pattern is performed promptly, precisely, and smoothly.

18-19: Generally excellent performance with one minor fault in appearance and position of exhibitor or execution of the pattern (performance).

16-17: Generally good pattern execution and equitation with one minor fault in precision or execution of pattern (performance), or appearance and position of exhibitor.

14-15: Average pattern that lacks quickness or precision, or rider has obvious equitation flaws that prevent effective equitation, or commits two or three minor flaws in the performance or appearance and position of exhibitor.

12-13: One major fault or several minor faults in the performance and/or appearance and position of exhibitor that precludes effective communication with the horse.

10-11: Two major faults or many minor faults in the performance or appearance and position of the exhibitor.

6-9: Several major faults or one severe fault in the performance, or appearance and position of exhibitor. Exhibitor demonstrates a complete lack of riding ability or commits a severe fault in the performance or appearance and position of exhibitor.

1-5: Exhibitor commits one or more servere faults in the performance, or appearance and position of exhibitor, but does complete the class and avoids disqualification.

In huntseat equitation over fences classes, riders are judged on their ability to guide the horse over a course of jumps. In some classes, riders may come back into the ring and do rail work following everyone's completion of the course. Riders are judged on their position and effectiveness. Riders may be inclined slightly forward when jumping the course but their weight should still be centered over the middle of the saddle (similar to riding a posting trot or a hand gallop). The riders will get in a 2 point position over the jumps; this is a position where riders hold themselves off the horse's back to allow it to jump freely, figure 16-15. Again, the rider's weight should be centered over the middle of the saddle and the rider should not be interfering with the horse's ability to use his head and neck. Riders who are not centered over the horse in a correct 2 point position will have trouble balancing and holding the position and in turn have trouble controlling the horse upon landing, figure 16-16. Riders should smoothly move into and out of the 2 point position. They should not pull back on the reins as the horse is jumping or interfere in any way with the horse's ability to jump. Riders are also judged on their ability to make correct decisions to rate their horse's pace and approach the fence properly. If the horse jumps from too close

FIGURE 16–15 A rider in correct 2 point position for jumping a fence. *(Courtesy of UGA NCAA Equestrian Team.)*

FIGURE 16–16 This rider is having trouble balancing in the 2 point position. She has almost no weight in her stirrups and is having trouble holding herself out of the saddle. These position flaws make the rider more likely to fall back into the saddle before the horse finishes jumping, thereby interfering with the horse's jumping ability. *(Courtesy of UGA NCAA Equestrian Team.)*

or too far away from the jump, the rider is faulted. The judge will evaluate the rider's effectiveness in handling the horse, getting it to the jumps properly, and presenting a smooth and effective ride.

WESTERN DISCIPLINES
Western Pleasure

Western pleasure horses are judged on their ability to produce a pleasurable ride. They should be obedient, have soft and smooth gaits, have a quiet and easy temperament, and be able to be ridden with little restraint. The class is judged at the walk, jog, and lope in each direction. A jog is a slower trot (still two beats) than a hunter under saddle trot and a lope is a slower canter (still three beats). The horse should carry himself in a quiet and relaxed frame with the head approximately vertical from poll to nose and his poll approximately level with his withers, figure 16-17. The horse may also be asked to extend its jog and should do so willingly without breaking gait (i.e., picking up the canter instead). Horses are also penalized if their gait is not true; for example, if the horse is supposed to be jogging but is actually walking on its hind end and trotting on his front end, this is penalized. The horse's gait should be correct; a jog should have two beats and a canter should have three. Riders use only one hand on the reins and are faulted for switching hands, figure 16-18. Riders may use two hands if showing in a snaffle bit or hackamore. This is typically only allowed for horses five years old and younger. When using one rein, riders must hold their reins with only one finger between reins. Riders are penalized for having more than one finger between the reins, touching the horse or saddle with their free hand, or spurring ahead of the girth. Horses are penalized for excessive speed or slowness, being on the wrong lead, breaking gait, failing to take the gait when called for, head being too high or too low, having the nose out in front of vertical or flexed behind vertical, stumbling or falling, opening the mouth excessively, and excessive head movement *AQHA Competitive Horse Judging Manual*, figure 16-19.

FIGURE 16–17 A horse in a correct frame for a western pleasure class. *(Courtesy of UGA NCAA Equestrian Team.)*

FIGURE 16–18 Rider showing in a western pleasure class using only one hand on the reins. Notice that the free hand is held in a position similar to the rein hand. *(Courtesy of UGA NCAA Equestrian Team.)*

FIGURE 16–19 This horse is resisting the bit by both opening its mouth and lifting its head. This action would be faulted in a western pleasure class. *(Courtesy of UGA NCAA Equestrian Team.)*

Horsemanship

Horsemanship is similar to equitation in that it is also judged on the rider's ability to ride and control a horse. The proper position for horsemanship is similar to that of the huntseat equitation rider. A straight line should be able to be drawn from ear to shoulder to hip to heel, figure 16-20. Stirrups length is somewhat longer than that of the huntseat rider. For western horsemanship, the stirrup should be just short enough for the heel to be lower than the toe. Stirrups should be placed on the ball of the foot; riders who have only their toe in the stirrup are penalized. The hand position, as described by the *AQHA Handbook,* is to hold both hands and arms in a relaxed manner with the upper arms in a straight line with the body, figure 16-21.

Riders ride with one hand on the reins and one finger between the two reins. Again, riding with two hands is permissible on young horses if the rider is using a snaffle, bosal, or hackamore. Hand position of the rider's free hand (the one not holding the reins) has varied based on trends in the industry. Currently the free

FIGURE 16–20 This rider has correct position for a western horsemanship class. A straight line can be drawn from her ear to shoulder to hip to heel. *(Courtesy of UGA NCAA Equestrian Team.)*

FIGURE 16–21 Correct arm position for a western horsemanship class. Note that the elbow is in line with the rider's shoulder. *(Courtesy of UGA NCAA Equestrian Team.)*

hand is generally held in a position similar to the hand that is holding the reins. Sometimes, one might see the free hand hanging by the rider's side and resting lightly on his or her leg. Riders should never change rein hands. Reins are carried either directly above or slightly in front of the saddle horn. Rein length should be long enough so that there is some slack in the reins and the rider does not appear to be pulling on the horse but short enough that the rider does not have to move the hand excessively to get a response. Favoritism is given to riders who have invisible aids, that is, those riders who can maneuver the horse without the judge seeing the cues the rider has given. Riders should sit in the saddle at all times; they should not post to the jog. A test or pattern will be given by the judge to be performed by all riders. This test will be scored on a 1-20 scale with the same guidelines as the huntseat equitation pattern. The horse/rider pairs will also be required to work on the rail; rail work is used to break ties or extremely close scores.

Reining

Reining is a western performance event that is judged on the horse's obedience and ability to maneuver easily. The horse/rider will perform one of ten patterns approved by the National Reining Horse Association (NHRA) of the AQHA. The specific pattern that is used at an event is selected by the judge. These patterns require that the horse be able to canter at different speeds both in circles and straight lines, perform sliding stops, back willingly, perform spins, and execute lead changes.

The philosophy of reining is that the rider controls every movement of the horse and the horse is willingly guided with virtually no resistance. Maneuvers that may be asked of the reining horse are as follows:

Sliding Stops

The horse is required to stop by dropping his hindquarters low to the ground while front end continues to move, thereby allowing the horse to slide to a stop. The stop should look controlled and smooth. The horse should be round at the loin and should lower his head and neck into the stop, figure 16-22.

FIGURE 16–22 A horse correctly performing a sliding stop. *(Courtesy of UGA NCAA Equestrian Team.)*

FIGURE 16–23 A horse correctly performing a spin in a reining class. Notice that the horse has one hind leg planted in the ground and is crossing one front leg over the other as it spins. *(Courtesy of UGA NCAA Equestrian Team.)*

Spins

These are very quick pivots around one hind leg of the horse. The horse plants one hind leg in the ground and spins quickly around it. Ideally, the spins should be low to the ground and the horse's front legs should cross in front of each other as he spins, allowing the spin to be fluid. The horse should stop the spin in exactly the spot dictated by the pattern, figure 16-23.

Rollbacks

This is a combination of a sliding stop and a spin. The horse accelerates into the stop, drops his haunches in the ground and slides to a stop, then immediately spins back the other direction and lopes off. There is no hesitation between the stop and the spin back.

Circles

The horse is asked to perform circles, either large and fast or slow and small. When performing a large fast circle, the horse should move forward willingly on loose contact. He should look straight forward and run with speed, figure 16-24. The horse should easily and willingly slow down into a small slow circle. In the small slow circle, the horse should be soft and collected, that is, willingly carry himself on a shorter stride with his head straight out in front of him flexed at the poll and held about wither height, figure 16-25.

Lead Changes

All lead changes in the pattern should be flying lead changes. The horse should change leads with its front and back legs at the same time and should be accurate. The horse will be penalized for changing after the point the pattern dictates. Ideally, the lead change will be flat and smooth, that is, the horse will not jump up into the lead change.

Throughout all of the maneuvers, the horse should be guided willingly and easily on a light rein. There should be absolutely no resistance to the rider's commands. The horse should be aggressive in his maneuvers but perfectly manageable

FIGURE 16-24 A horse performing a large, fast circle in a reining pattern. *(Courtesy of UGA NCAA Equestrian Team.)*

FIGURE 16-25 A horse performing a small, slow circle in a reining pattern. *(Courtesy of UGA NCAA Equestrian Team.)*

by the rider. The horse should not be fussy with the bit or his tail (e.g., no excessive mouthing of the bit or swishing of the tail).

Reining patterns are scored on a scale of 1-100 with 70 representing an average score. Points are deducted at very specific levels for specified disobediences. Points can also be awarded if a horse performs a maneuver that is above average. According to the *AQHA 2007 Official Handbook,* the following will result in disqualification and no score being given.

Abuse of the horse in the show arena or evidence of previous abuse:

> *Use of illegal equipment*
>
> *Use of illegal bits*
>
> *Use of whips*
>
> *Failure to provide horse and equipment to the judge for inspection*
>
> *Disrespect by the exhibitor*
>
> *Use of closed reins other than standard romal reins*

The following incidences will result in a score of 0 being given:

> *Failure to complete the pattern as written*
>
> *Performing maneuvers other than in a specified order, including maneuvers not specified (e.g., backing when not supposed to, spinning more than 90 degrees where not specified)*
>
> *Equipment failure resulting in a delay of completion of the pattern*
>
> *Jogging in excess of half of a circle or half the length of the arena when circling, starting a circle or exiting a rollback*
>
> *Spinning more than ¼ of a turn past what is specified*
>
> *Use of a romal other than in the way indicated in the handbook*
>
> *More than one finger between the reins*
>
> *Using two hands on the reins unless riding a young horse in a snaffle bit or hackamore*
>
> *Changing hands on the reins*
>
> *Fall to the ground of horse or rider*
>
> *Balking or refusing a command*

Some faults are considered very serious but do not result in elimination. The following faults result in a 5 point deduction:

Spurring in front of the girth

Using the free hand to instill fear or praise in the horse

Holding the saddle with either hand

Any blatant disobedience results in 5 points being deducted per incident (e.g., kicking out, biting, bucking, rearing).

The following faults are more minor but are also specified and result in a 2 point deduction per incident:

Failure to go past the markers on stops and rollbacks

Breaking of gait

Freezing up in a spin or rollback

When using walk-in patterns, failure to stop or walk before executing a canter departure

On run patterns, failure to be in a canter before the first marker

Other penalties are also defined in the *AQHA Handbook*. Each time a horse is on the wrong lead, 1 point is deducted. Also, when performing circles, the judge is required to deduct 1 point for each quarter of the circumference of the circle or any part thereof that the horse is on the incorrect lead. The judge also must deduct 0.5 point for a change of lead that is delayed by one stride. In addition, if the horse jogs up to two steps at the beginning of a circle or when exiting a rollback, 0.5 point is deducted each time. If the horse jogs more than two strides (but less than half the circle or length of the arena), 2 points are deducted. When spinning, there are a specified number of turns the horse must make. A 0.5 point deduction is enforced if the horse over or under spins by one-eighth or less of a turn. One point is deducted for over or under spinning by one-eighth or one-fourth of the turn. Also, the horse must remain at least 20 feet from the wall or fence of the arena when beginning a stop or rollback. If he does not allow this distance, 0.5 point is deducted. When a lead change is required prior to a run at the end of the arena (i.e., if the horse has been circling and then is required to exit the circle and run down to the end of the ring), points are deducted based on how far the horse goes before executing the lead change properly. If the horse goes one stride into the turn before changing, 0.5 point is deducted. If the horse does not change leads by one stride but does change before the next maneuver, 1 point is deducted. If the horse does not change leads before the next maneuver, 2 points are deducted. In cases where the horse is required to run around the end of the arena, 1 point is deducted if he is not on the correct lead when moving into the turn around the end of the arena. If he has not completed the lead change by the center of the end of the arena, 2 points are deducted. For example, if the horse runs down one end of the ring and stops, then performs a rollback turn and runs back in the direction he just came from, it does not matter what lead the horse picks up out of the rollback. However, if he is required to turn left at the end of the arena, he must change to the left lead before he turns.

Other faults that are penalized at the judge's discretion are the following:

Opening the mouth excessively; mouthing the bit

Opening the mouth or raising the head significantly when performing sliding stops, figure 16-26

FIGURE 16–26 A horse raising its head and opening its mouth while performing the sliding stop. *(Courtesy of UGA NCAA Equestrian Team.)*

FIGURE 16–27 A horse stiffening its legs and raising its head, leading to a bouncy stop. *(Courtesy of UGA NCAA Equestrian Team.)*

> *Stops that are bouncy or are not straight, figure 16-27*
>
> *Refusing to change leads*
>
> *Anticipating signals and performing maneuvers before asked*
>
> *Stumbling*
>
> *Failure to back in a straight line*
>
> *Knocking over any markers in the arena*

In addition, the rider can also be penalized for the following actions:

> *Feet out of the stirrups (called losing a stirrup)*
>
> *Failure to go beyond markers on rollbacks and stops*

In addition to penalty points, quality points are added or deducted as well. These points are awarded or subtracted based on the judge's opinion of how the horse performed each maneuver. A NHRA judge's score card is broken down into a series of different maneuvers. Each maneuver is assigned quality points ranging from +1.5 to −1.5. The scoring breakdown is as follows:

+1.5	Excellent
+1	Very good
+0.5	Good
0	Correct
−0.5	Poor
−1	Very poor
−1.5	Extremely poor

For a total score, quality points and penalty points are totaled together and either added to or subtracted from 70. The horse with the highest score is the winner.

Western Riding

Western riding is a unique event where the horse is judged on its movement, obedience, and ability to perform technical maneuvers. The horse is required to perform a specified pattern that includes flying lead changes, agile turning, and negotiating over a log. According to the *AQHA Official Handbook,* the horse is judged on its quality of gaits, smoothness, precision and quality of lead changes, obedience to the rider, and willingness of attitude. It is expected that the horse perform at a reasonable speed, have evenly cadenced gaits (start and finish the patterns with the same cadence), change leads correctly and simultaneously on the front and hind legs with precision, have a relaxed head carriage and flex moderately at the poll, and respond easily to imperceptible cues from the rider. The horse should not look like it is rushing through the pattern but should complete each maneuver promptly. The horse should cross the log without breaking gait or touching the log. This is a scored event with the scale ranging from 0–100 and 70 representing the average score. Each of eight required maneuvers is given a score from +1 to −1.

The scores are assigned as follows:

+1   excellent

+0.5   good

0   average

−0.5   poor

21   very poor

A typical pattern is shown in figure 16-28, and the nine maneuvers that are scored are the following: (1) gate, (2) walk/log, (3) jog, transition to lope, (4) line (side) lead changes (+/− 1 point per change), (5) first two crossing lead changes (+/− 1 point per change), (6) log/lope, (7) second two crossing lead changes (+/− 1 point per change), (8) lope, stop, and back, (9) overall pattern accuracy and smoothness (+/− 1 point).

Penalties are deducted from the competitor's score for the following reasons:

Out of lead beyond the next designated lead change area:	−5 points
Blatant disobedience (kicking, biting, rearing, etc.):	−5 points
Not performing the specified gait or not stopping when called for within 10 feet of designated area:	−3 points

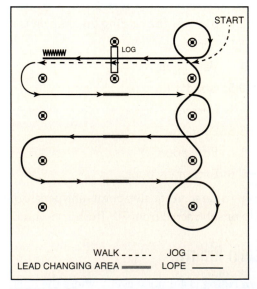

1. Walk, transition to jog, jog over log
2. Transition to the lope, on the left lead
3. First crossing change
4. Second crossing change
5. Third crossing change
6. Circle and first line change
7. Second line change
8. Third line change
9. Fourth line change and circle
10. Lope over log
11. Lope, stop, and back

FIGURE 16–28 An example of a former western riding pattern.

Breaking gait at the lope (i.e., performing a simple change instead of a flying change):	−3 points
Out of lead at or before the marker prior to the designated change area or after the marker after the designated change area:	−3 points
Performing lead changes where not required:	−3 points
Failure to pick up the required gait at the point designated on the pattern:	−3 points
Breaking gait at the walk or trot:	−3 points
Hitting the log:	−1 point
Out of lead by one stride on either side of the center point and between markers:	−1 point
Touching the log lightly:	−0.5 point
Failure to change leads immediately behind (i.e., cross cantering—changing leads on the front end but not the hind end):	−0.5 point

A horse is disqualified if the rider goes off pattern, knocks over any markers, misses the log, uses illegal equipment, abuses the horse, has a major refusal or disobedience, performs four or more simple lead changes, or overturns more than one-fourth of a turn. Other occurrences that are penalized at the judge's discretion are (1) excessive mouthing of the bit or opening of the mouth, (2) anticipating the next signal or changing leads before the pattern requires it, (3) stumbling of the horse, (4) head carried too high or low, (5) overflexing by the horse or excessive nosing out. All penalties and quality points are totaled and either added or subtracted from 70 for a final score. The horse with the highest score is the winner.

GAITED HORSES/SADDLE TYPE HORSES

Gaited and saddle type horses are shown in a style that originated from plantation days. Many of these breeds will show both English and western but are different from the classes previously discussed because of their distinctive movement and gaits. Breeds that can be put into this group include the American Saddlebred, the Morgan, the Arabian, the Tennessee Walker, and the Paso Fino. Typically, Morgans and Arabians have similar gaits to those described above; that is, they perform at the walk, trot, and canter. They will often show in both English and western pleasure, equitation, and other performance classes. English pleasure is different from hunter under saddle, and western pleasure is different from the above described western pleasure in the way the movement is judged. In general, Arabians and Morgans will move with more knee action than what is acceptable in the classes previously described. Arabians and Morgans will typically have slightly higher head and tail sets, may be slightly more flexed at the poll, and will lift their knees higher when moving when compared to Quarter Horses and other stock breeds, Warmbloods, and Thoroughbreds, figure 16-29.

FIGURE 16–29 An example of an Arabian horse. Notice that the Arabian has a higher head and tail set as well as more knee and hock action when compared to Quarter Horses, Thoroughbreds, and Warmblood breeds. *(Delmar/Cengage Learning. Photo by Kylee Duberstein.)*

Other gaited and saddle type breeds may have gaits which are unique to them. Other gaits that might be encountered at certain breed shows include running walk, foxtrot, pace, slow gait, rack, and paso. Some of these gaits are natural gaits for certain breeds of horses and some of the gaits are artificially created or enhanced.

The running walk is a gait that is natural for the Tennessee Walker. It is similar to a walk in that it has the same footfalls, but in this gait the hind leg oversteps the front leg track by several inches. It is significantly faster than a normal flat footed walk (walk is approximately 4 miles per hour while running walk is approximately 6.5 miles per hour). The running walk is easy for the rider to sit but is a strenuous gait for the horse to maintain.

The foxtrot is similar to the trot but is a four-beat gait rather than a two-beat gait. It is a slow and short trot where the hind leg will hit the ground just before the front leg of a diagonal pair. This gait is natural for the Missouri Fox Trotter.

A pace is similar to a trot in that it is a two-beat gait but when pacing, the horse moves its legs in lateral pairs rather than diagonal pairs (i.e., the horse moves its front and hind leg on the left side forward simultaneously and then moves the front and hind leg on the right side forward simultaneously). This is somewhat of a natural gait for some breeds. It is commonly used as a gait in harness racing (Standardbred horses).

A slow gait is an artificial gait performed by the American Saddlebred. It is a slow, broken pace that is four beats where the hind leg on one side lands slightly before the front leg on the same side followed by the hind leg of the other side and then the front leg of that side. It is performed slowly with high knee action and more than one leg can be on the ground at one time. It is approximately the same speed of a walk (4 to 5 mph). An extended version of the slow gait is the rack. This is an artificial four-beat gait that is performed so quickly that only one foot touches the ground at a time (typically is 12 to 13 mph). This gait is often times performed by horses of Tennessee Walking Horse origin or is the fifth gait of the five gaited American Saddlebred show horses.

Finally, the paso is a natural four-beat gait found in some Latin American breeds such as Paso Finos and Peruvian Pasos. It has a similar foot fall to the pace and the slow gait. Three feet are generally on the ground at all times. The horse moves its hind leg followed by its front leg on one side and then its hind leg followed by its front leg on the other side. It is moderately slow with high knee action. The speed may vary greatly (2 to 9 mph).

There are many different classes and events that showcase the different breeds and different gaits within a breed. Judging in most all classes is based on quality of movement and obedience and manners of the horse. It would be impossible in the scope of one chapter to go into great detail of each class offered. Each breed association and horse show association will go into great detail on class specifications and judging criteria. Typically this information can be found in the rulebook for each organization.

SUMMARY

Judging riding classes requires that the judge rank a group of horses based on a criteria set forth for a particular discipline of riding and specific class within that discipline. Riding disciplines can be divided into two basic groups: English and western. These two main groups can be further divided into a multitude of performance events. Furthermore, there are numerous breed organizations that host events to showcase the breed of horse they represent. Horse gaits and carriage can vary from breed to breed and between styles of riding. It is critical that you understand specific rules for all classes offered at a horse show or horse judging competition in which you may be a participant.

STUDENT LEARNING ACTIVITIES

1. Watch either a real horse or a video of a horse and name the different gaits that the horse is performing as it moves.
2. Pretend you are a horse and demonstrate how you would move your legs while traveling on the right lead at a canter.
3. Look at both an English saddle and a western saddle and discuss how they might have originated to look as they do today based on their past and present purposes.

TRUE OR FALSE

T	F	1. Most performance events can be divided into two main disciplines of riding: Eastern events and western events.
T	F	2. English riding is traditionally considered forward seat riding.
T	F	3. A walk is a slow, three-beat gait.
T	F	4. A horse's right front leg is leading when the horse is on the right lead at the canter.
T	F	5. A flying lead change is when a horse breaks to a trot to change leads.
T	F	6. Equitation and horsemanship classes are judged on the rider.
T	F	7. Hunter under saddle classes are judged both on the rail and over two small jumps.

T   F   **8.** Western riding and reining are the same event.

T   F   **9.** A gaited horse may perform a running walk, a fox trot, a slow gait, a rack, or a paso.

T   F   **10.** The running walk is a gait which is natural for the Tennessee Walking Horse.

 ## FILL IN THE BLANKS

1. Typically when judging stock horse breeds (Quarter Horses, Paints, Appaloosas, etc.), and sport horse breeds (Thoroughbreds and Warmbloods), the four gaits seen are the _____, _____, _____, and the _____.

2. A _____ is a medium paced gait (approximately 4 m/s) that consists of two beats; the legs move in diagonal pairs.

3. The horse's _____ refers to the way the horse carries itself (head carriage, stride length, etc.).

4. In both hunter under saddle and western pleasure classes, the poll of the horse should be flexed to where a straight line can be drawn from the _____ to the _____ that is perpendicular with the ground.

5. The hunter hack class is somewhat of a combination of the _____ and _____ classes.

6. In equitation and horsemanship classes, a straight line should be able to be drawn from the rider's _____ to _____ to _____.

7. In western pleasure classes, a _____ is a slower trot than in a hunter class (still two beats) and a _____ is a slower canter (still three beats).

8. When showing in a western event, a rider may only use two hands on the reins if showing in a _____ or _____.

9. The philosophy of reining is that the rider controls every movement of the horse and the horse is willingly guided with virtually no _____.

10. A _____ is similar to a trot in that it is a two-beat gait but when pacing, the horse moves its legs in lateral pairs rather than diagonal pairs.

 ## DISCUSSION

1. Describe the criteria for which a hunter under saddle is judged.

2. Describe jumping form that would be considered unsafe.

3. Why does take-off distance at a jump affect the jump or the performance of the horse?

4. Describe the frame a western pleasure horse should have.

5. How should an equitation rider's hands be positioned?

6. How should a horsemanship rider carry his or her free hand?

7. Describe how a reining horse may be penalized on its sliding stops.

8. Describe at least four maneuvers a western riding horse might be judged on.

9. Describe knee action in regard to a Quarter Horse and an Arabian.

10. Describe the footfall pattern of a running walk and a foxtrot.

APPENDIX I

In Appendix I are diagrams with the parts of cattle, sheep and swine labeled. The anatomy and terms used for horses are covered in chapter 14. Be sure you can correctly identify all the parts before you continue with learning to evaluate livestock. Terms are courtesy of Mark Spangler and are used with permission.

PARTS OF CATTLE

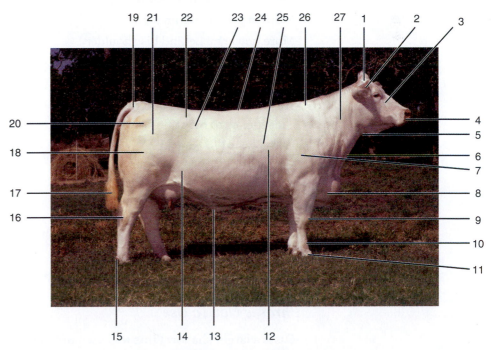

(Courtesy of American – International - Charlois Association.)

1. **Poll**	2. **Ear**	3. **Face**
4. **Muzzle**	5. **Throat**	6. **Dewlap**
7. **Point of Shoulder**	8. **Brisket**	9. **Knee**
10. **Pastern**	11. **Hoof**	12. **Heart Girth**
13. **Belly**	14. **Flank**	15. **Dew Claw**
16. **Hock**	17. **Switch**	18. **Quarter or round**
19. **Tail Head**	20. **Pins**	21. **Rump**
22. **Hooks**	23. **Loin**	24. **Back**
25. **Ribs**	26. **Crops**	27. **Neck**

Cattle terms

Structure

-Sounder off of both ends
-More correct angle to his shoulder
-Longer more accurate stride
-More flex and give to his hock and pastern
-Takes a longer stride off of both ends
-Stood squarer from behind

-Straight off of both ends
-Straight in his shoulder
-Short stride
-Straight in his hock and pastern

-Cow-hocked

Muscle

-Bigger topped and thicker ended
-More powerful from behind
-More impressive from behind
-More powerful behind his shoulder
-Thicker hip and more expressive quarter
-More width to pins and deeper quarter
-Thicker out of his hip and descends this deeper onto his stifle
-Wider based or gauged
-More volume of muscle

-Narrow topped and flat thru his quarter

-Narrows out of his hip

-Narrow based or gauged

Overall Appearance

-More eye appealing
-Simply more of him/her
-The most complete, tying together a better combination of ......

Phrases Not To Use

-Casts a larger shadow (This was used decades ago when being extremely large framed was considered good)
-Larger framed (This is not always a good thing, use with caution.)

-Bigger outlined (This is not always a good thing, use with caution)

-More pounds of product (There are more descriptive terms to use in its place)

Breeding cattle specific terms

Femininity and Balance

-Cleaner jointed or flatter boned

-More refined about her head and neck

-Longer, thinner neck

-Cleaner patterned/conditioned

-Lays her tail head in more correctly
  at the end of her spine

-Longer, leveler top

-Straighter in his lines

-Lays in smoother at his shoulders

-Neater dewlap

-Course jointed

-Short, thick neck

-Heavy conditioned

-Course about her tail head

-Short, weak top

-Course at the point of his
  shoulders

-Leathery fronted

Body Dimension

-Deeper sided/bodied/ribbed

-Bolder sprung

-Bigger bodied

-Broodier appearing

-Appears to be easier fleshing

-Strikes me with more of a brood cow look

-More brood cow potential

-Spreads more volume of muscle down
  her top

-More uniform in her body depth

-Shallow sided/bodied/ribbed

-Flat ribbed/narrow chested

-Appears to be hard keeping

-Pinched in her heart or shallow
  in her flank

Masculinity

-More testicular development

-His testicles descend more correctly
  from his body

-More correct in his testicular carriage

-Small testicles

-Twisted scrotum

Market steer specific terms

Finish

-Cleaner thru his chest and middle

-Deeper in his flank

-Fuller in his fat indicators

-Deep chested, heavy middled

-Shallow flanked

-Handles more uniform in his cover	-Handles patchy in his cover
-Handles with more finish over his fore and rear rib	-Harsh handling, or bare

Body Dimension

-Deeper sided/bodied/ribbed	-Shallow sided/bodied/ribbed
-Bolder sprung	-Flat ribbed/narrow chested
-Bigger bodied	
-More uniform in his body depth	-Pinched in his heart or shallow in his flank

Balance

-Lays his tail head in more correctly at the end of her spine	-Course about his tail head
-Longer, leveler top	-Short, weak top
-Straighter in his lines	
-Lays in smoother at his shoulders	-Course at the point of his shoulders

Carcass

-He is heavier muscled and should open up with a larger ribeye area

-His leanness should give him a yield grade advantage

-He is the heavier muscled, leaner steer who should have a higher cutability carcass

-He is the bigger topped thicker ended steer who should rib with a larger eye

-He is cleaner through his chest and lower shoulder, but still adequate in terms of finish. Thus I would expect him to have a more desirable combination of quality and yield grade.

-He should open up a larger ribeye, exposing a higher degree of marbling.

-He is the cleaner patterned steer that handles leaner over his ribs and loin edge and thus should kill with a yield grade/cutability (if heavy muscled too) advantage.

-He is the more moderately framed steer who meets my hand with a greater turn to his loin edge and thus should kill with more ribeye relative to his carcass weight.

-He is the more moderately framed steer that is easily bigger topped and thicker ended and still comparable in terms of leanness. Thus his smaller, more muscular carcass should create a yield grade advantage.

-He is the most complete steer as he not only excels on the hoof, but he should also prove to be the best (or most valuable) in the cooler.

-He is the more product oriented steer who is bigger topped and wider based.

-Yes, he appears to have been the easiest feeding, but today this works to his disadvantage, as he is the most excessive in his cover and should have the least desirable yield grade of the class.

-Yes, he should have the lowest yield grade, but this goes with his kind. He is also the tightest ribbed and appears to have been the hardest feeding.

-He is closer to optimum in his degree of finish and this coupled with his Angus influence suggests he is safer into a higher quality grade (or the choice grade).

PARTS OF A SHEEP

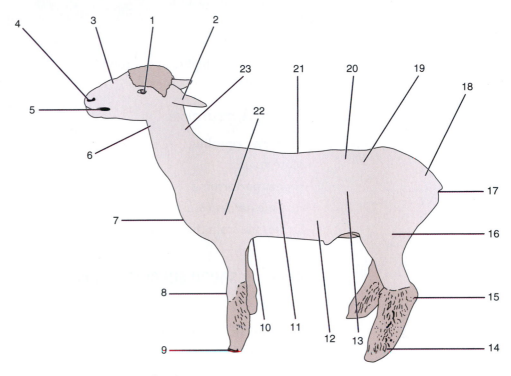

(Delmar/Cengage Learning.)

1. Eye	2. Ear	3. Face
4. Nose	5. Muzzle	6. Throat
7. Breast	8. Knee	9. Toe
10. Heart Girth	11. Ribs	12. Middle
13. Flank	14. Pastern	15. Hock
16. Leg	17. Dock	18. Rump
19. Edge of Loin	20. Loin	21. Back
22. Shoulder	23. Neck	

Sheep terms

Frame

-Larger framed	-Small framed
-Bigger outlined	
-Heavier weight	-Light weight
-Growthier	-Early maturing
-Taller fronted	-Low fronted
-More skeletal extension	-Short sided or short coupled

Balance

-Has a longer thinner neck	-Short, thick neck
-Ties a longer, thinner neck into a smoother shoulder	
-Lays in tighter at the top of his shoulder	-Open and course shouldered

-Longer and leveler from the top of his
  shoulders to his dock

-Leveler hipped   -Steep hipped

-Tighter hided   -Pelty

-Ties his neck in higher at the top of   -Ewe necked/low at neck
  his shoulders   shoulder junction

Overall Appearance

-More eye appealing

-More stylish

-Higher performing

-Hits me harder from the side

-Strikes me from the profile

Breeding sheep specific terms

Muscle

-Stouter made and heavier muscled   -Frail and light muscled

-Is thicker and fuller out to his/her dock -Tapers out to his dock

-More muscle volume

-More natural thickness

-Wider based or tracking   -Narrow based or tracking

Body Dimension

-Bolder sprung   -Flat sided

-More body dimension

-More body capacity   -Tight ribbed

Structure

-More correct on his/her feet and legs

-Stronger pasterns   -Weak pasterns (slang is coon
                                           footed)

-Stood squarer from hock to ground   -In at his/her hocks

Market lamb specific terms

Finish

-Cleaner patterned

-Cleaner or trimmer thru his breast   -Wasty thru his breast
  and middle   and middle

-Handles leaner over his fore rib   -Soft over his fore rib

-Handles firmer down his top

Muscle

-Stouter made and heavier muscled	-Frail and light muscled
-Handles with more tone and expression to his rack/loin edge/inner leg	
-Handles with a larger leg circumference	
-Handles with a plumper, meatier leg	
-Spans the wider, deeper loin	-Narrow and shallow over his loin
-Is thicker and fuller out to his dock	-Tapers out to his dock
-Handles with a thicker, fuller loin edge	
-Handles fresher over his rack and loin	-Pinched over his rack and shelly thru his loin
-Measures longer from his last rib back or with the longer hind saddle or loin or hip	-Measures with a short hind saddle

Carcass

-Handles with more product from his last rib back.

-He is cleaner patterned and handles leaner over his fore rib; thus he should have an advantage in yield grade.

-Higher cutability (heavier muscled and leaner).

-He is the leaner handling lamb who meets my hand with more tone and expression to his rack and inner leg. Consequently, he should hang a shapelier carcass.

-He simply handles with more meat animal shape.

-He is the biggest outlined, cleanest patterned lamb and with his kind comes a cutability advantage.

-He is the wider gauged more product driven lamb who has a weight advantage and spreads more muscle from his last rib back (or spans the wider top and descends into a larger leg).

-A shapelier carcass with a higher conformation score.

Wool Terms

Wool terminology should only be used when describing wool type sheep when the contestant is allowed to handle the fleece.

Britch Wool:	Usually the coarsest in the fleece, from the lower parts of the hindquarters.
Burry Wool:	Wool that contains burrs from any plant source that will require special processing.
Character:	The evenness and distinctness of crimp in wool fibers.
Clip:	The process of shearing or clipping. The weight or type of wool from a certain flock.
Condition:	Refers to the amount of dirt or grease in the fleece. A fleece that is heavy conditioned will shrink or lose a larger percentage of its weight in scouring.

Cotted Fleece:	Fleece in which the fibers are matted or felted.
Crimp:	The natural waviness in the fibers. A tighter crimp is more desirable
Grease wool:	Shorn wool before washing or scouring.
Kemp:	A chalky white, weak, brittle fiber bound mixed with normal fibers of a fleece. Kemp will not take dyes.
Lanolin:	Wool grease or yolk that has been refined.
Luster Wool:	Wool that shines because it reflects more light.
Quality:	A term used to describe the fineness of the fibers.
Raw Wool:	Wool in the Grease.
Run-Out Fleece:	A Fleece that lacks uniformity, being hairy or kempy in the britch or elsewhere.
Scouring:	Removing the grease and dirt from the wool.
Spinning count:	A numerical system of wool grading on the basis of the number of hanks of yarn that can be spun from it.
Staple:	Used in reference to length of wool fiber. Longer is more desirable.
Tags:	The heavy manure-covered locks.
Yield:	The percentage of clean wool fibers after scouring.
Yolk:	The secretion of sebaceous or oil glands in the skin. A certain amount is needed to keep the wool in good condition.

PARTS OF A PIG

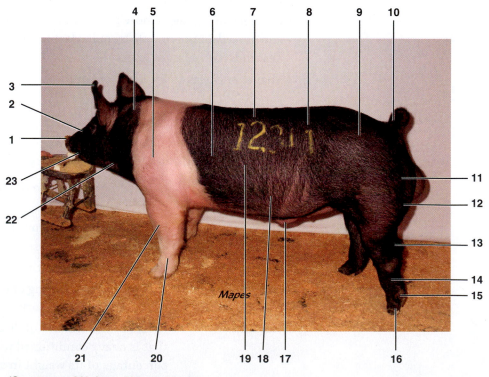

(Courtesy of Heimer Hampshires.)

1. Snout	2. Face	3. Ear
4. Neck	5. Shoulder	6. Ribs
7. Back	8. Loin	9. Rump
10. Tail	11. Ham	12. Cushion of Ham
13. Hock	14. Cannon	15. Dewclaw
16. Toe	17. Sheath	18. Belly
19. Side	20. Pastern	21. Knee
22. Jowl	23. Mouth	

Swine terms

Muscle

-Built wider from the ground up

-Pulled apart wider at his/her blades

-More opened up in his/her top

-Has a bigger, squarer loin laying down his/her top

-More powerful ham or from behind

-More muscular turn to his/her loin edge

-Spreads more volume of muscle from blade to hip

-Deeper tying ham

-More honest turn to his loin edge

-A butterfly shape top

-Narrow based

-Pinched in his/her blade

-Non-descript in his top

Leanness

-Cleaner constructed

-Cleaner thru his/her jowl and lower body

-Totally freer of fat throughout

-Cleaner profiling

- He/She reads leaner at his/her 10th

-Wasty thru his jowl and lower body

-Reads fatter at his/her 10th

Structure

-More mobile

-Works out of a longer, looser hip

-More set back at his/her knee

-More correct in the angle to his/her shoulder

-Moves with more flex and give to his/her hock

-Takes to the ring with more mobility out of both ends

-More durable

-Short and tight hip

-Over at his/her knee

-Straight/upright in his/her shoulder

-Straight in his/her hock and pastern

-Frail

-Moved with more stability

-More even toe size   -Small inside toe

-Truer tracking from the rear   -Hocks in

-More confinement adaptable

Breeding swine specific terms
Underlines

-More refined about her underline   -Course about her underline

-More prominent underline   -Blunt in her underline

-More uniform in her teat size and placement

-More functional underline   -Pin nipples

Market hog specific terms
Carcass

-Should generate more premiums on a lean value based system

-Has more potential lean cut out value

-Should rail the higher cutability carcass

-Simply spreads more product from blade to hip

-Should rank higher on a fat-free lean index

Appendix I is courtesy of Dr. Matt Spangler

APPENDIX 2

I. OBJECTIVES

1. To understand and to interpret the value of performance data based on industry standards.
2. To measure the students' knowledge in the following categories:
   a. to make accurate observations of livestock
   b. to determine the desirable traits in animals
   c. to make logical decisions based on these observations
   d. to discuss and to defend their decisions for their placing
   e. to instill an appreciation for desirable selection, management and marketing techniques
3. To develop the ability to select and market livestock that will satisfy consumer demands and provide increased economic returns to producers. Provide positive economic returns to producers as well as meet the needs of the industry.
4. To become proficient in communicating in the terminology of the industry and the consumer.
5. To identify the criteria used in grading livestock. Scenarios will be used in the selection process.
6. To provide an opportunity for participants to become acquainted with professionals in the industry.

II. EVENT RULES

1. Participants will report to the event superintendent for instructions at the time and place shown in the current year's team orientation packet.
2. Data may be added or deleted as technology changes. When new criteria are adopted, the information will be forwarded to all states by January 1 of the event year by the National FFA Program Manager responsible for Career Development Events.

III. EVENT FORMAT
A. Equipment

Materials students must provide- Participants must bring two (No. 2) pencils.

Equipment provided- Participants are not to bring any paper or clipboards. All paper and support boards will be provided.

B. Team Activity (*50 points/class 150 points total*)

Keep/cull classes- There will be three female selection classes, one each in beef, sheep and hogs, made up of eight animals. Participants will be required to select the four best animals from the eight, using visual appraisal and performance data. Performance data will be provided. All three of the female selection classes will make up the cooperative team activity.

Performance Records (including EPD's) may be used in the breeding and the keep/cull classes of beef, sheep and swine. Performance criteria, when used, shall be based on standards developed and used by the Beef Improvement Federation, the Sheep Industry Development Program, Inc. and the National Swine Improvement Federation. Participants will be allowed 15 minutes for each keep/cull class.

C. Individual Activities

1. **Livestock classes-** Six classes of livestock of four animals each will be placed using a computerized scorecard. There will be one class each of breeding and market beef, sheep and swine. Participants will be allowed 15 minutes for each class. (50 points/class 300 points total)

2. **Oral reasons-** Four sets of oral reasons will be designated by the event superintendent at the beginning of the event. One set of reasons will be given on the production data class. Reasons will be given after all classes have been placed. Notes will not be permitted; however, participants may use a card with only their placing of the class written on it. (50 points/class 200 points total)

3. **Beef slaughter class-** One class of 5 slaughter cattle will be graded individually, according to the latest U.S.D.A. market grades. Forms will be provided in the team orientation packets that are distributed annually prior to the event. The class will also be graded according to cutability. (50 points)

4. **Production Data Class-** One class of breeding beef, sheep or swine (4 animals) will be evaluated. Production data will be provided for each animal and will be utilized in the final placing of the class. This class will also be one of the four

oral reasons classes. Participants are expected to utilze the production data information provided as part of their oral reasons.

5. **Written test-** A multiple choice exam will be given. The objective exam is designed to determine team members understanding of the livestock industry. The exam will consist of 50 multiple choice questions. Sixty minutes will be given for the exam. (100 points)

IV. SCORING

	POSSIBLE POINTS	
	INDIVIDUAL	**TEAM**
Selection Classes		
Beef		
Placings [2]	100	300
Oral Reasons [1]	50	150
Sheep		
Placings [2]	100	300
Oral Reasons [1]	50	150
Swine		
Placings [2]	100	300
Oral Reasons [1]	50	150
Beef Slaughter Class	50	150
Production Data Class	50	150
Production Data Oral Reasons	50	150
Written Test	100	300
Team Activity (Keep/Cull)		150
Total Team Score**	700	2250

**(top 3 individual's scores plus Team Activity)

V. TIEBREAKERS

If ties occur, the following events will be used in order to determine award recipients:

1. Total of oral reasons
2. Total of placing classes
3. Total of grading classes

VI. AWARDS

Awards will be presented at an awards ceremony. Awards are presented to teams as well as individuals based upon their rankings. Awards are sponsored by a cooperating industry sponsor(s) as a special project, and/or by the general fund of the National FFA Foundation.

The individual and the team scoring the highest in each species of livestock and oral reasons shall receive special recognition.

VII. REFERENCES

This list of references is not intended to be inclusive. Other sources may be utilized and teachers are encouraged to make use of the very best instructional materials available. The following list contains references that may prove helpful during event preparation.

- For the most current copies of U.S.D.A. market standards and posters (large and small) illustrating market grades, write:

  Agricultural Marketing Service, U.S.D.A.

  Livestock Seed Event-

  Standard and Review Branch

  Room 2641 South Building

  Washington, DC 20250

- For the most current copies of performance criteria standards write:

Beef Improvement Federation

Department of Animal Sciences and Food

Kansas State University

Northwest Research Extension Center

105 Experiment Farm Road

Colby, KS 67701

National Swine Improvement Federation

204 Polk Hall

North Carolina State University

Raleigh, NC 27695-7621

Gillispie, James R. *Modern Livestock and Poultry Production*. 7th Edition. Albany, NY: Delmar Publishers, Inc. 2004.

Hunsley, R. *Livestock Judging, Selection and Evaluation*. 5th Edition. Danville, IL: Interstate Publishers, 2001.

Agriscience 332 – Animal Science (8831B). Instructional Materials Service, Texas A & M University 2588 TAMUS, College Station TX 77843-2588 (979)845-6601

Agriscience 231 – Plant and Animal Production (8651B). Instructional Materials Service, Texas A & M University, 2588 TAMUS, College Station TX 77843-2588 (979)845-6601

VIII. EXAMPLE

Keep/Cull Class Sample Card

Beef/Sheep/Swine

Participant/Team Name: _____

Participant/Team No.:_____

Circle or list the numbers of the 4 animals you want to keep as replacements.

1 - 2 - 3 - 4 - 5 - 6 - 7 - 8

Event officials will assign a point value to each one of the individual animals, giving the most points to the most desirable animal and the least points to the least desirable animal. If the participant selects the best four animals, full credit will be given.

OFFICIAL PLACING-KEEP/CULL

8	7	6	5	4	3	2	1	Animal
18	13	11	8	7	4	3	0	Points

KEEP/CULL CLASS SCORES

Participant									Score
A	**6**	11	**3**	4	**4**	7	**2**	3	**25**
B	**8**	18	**7**	13	**6**	11	**5**	8	**50**
C	**7**	13	**6**	11	**5**	8	**1**	0	**32**

Animals selected are shown in regular font with point values for that particular
animal shown in bold font. Point values are established by official judges
and will differ with each class.

Keep/Cull Team Activity Scorecard

Name:_____   Chapter:_____

State:_____   Team No.:_____

Member No.:_____

KEEP/CULL TEAM ACTIVITY

BEEF/SHEEP/SWINE – (SELECT ONLY ONE SPECIES)
Circle or list the numbers of the 4 animals you want to keep as replacements.

1 - 2 - 3 - 4 - 5 - 6 - 7 - 8

Event officials will assign a point value to each one of the individual animals, giving the most
points to the most desirable animal and the least points to the least desirable animal. If the participant selects the best four animals, full credit will be given.

OFFICIAL PLACING-KEEP/CULL

8	7	6	5	4	3	2	1	Animal
18	13	11	8	7	4	3	0	Points

Participant									Score
A	**6**	11	**3**	4	**4**	7	**2**	3	**25**
B	**8**	18	**7**	13	**6**	11	**5**	8	**50**
C	**7**	13	**6**	11	**5**	8	**1**	0	**32**

Animals selected are shown in regular font with point values for that particular animal
shown in bold font. Point values are established by official judges and will differ
with each class.

Judge's   _Judge's Signature_   _Date_

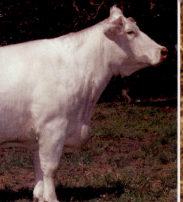

GLOSSARY

21 Day Litter Weight (LWT)—21-Day Litter Weight refers to the weight of the litter of pigs at the wean date. The industry standard & recommended days for weaning pigs is 21 days.

A

accuracy—a concise ability to correctly evaluate an animal. Accuracy is important in livestock judging as it allows participants to precisely evaluate an animal and explain their reasoning in an accurate manner.

adjusted days to 250 pounds—a number give to an animal that refers to the days it will be on feed before reaching 250 pounds. An ideal market hog will weigh around 250 pounds when it is taken to the processing facility. This uniform weight allows packers the ability to estimate date for loin eye area.

adjusted yearling weight—a set of adjustment factors that have been calculated to help compare yearling weights from different lambs on a more even basis.

adjusted—livestock number and data calculated to try and all the data to a relatively common value or average. This data includes many variables that might affect the numbers give for an individual animal.

aitch bone—the split portion of the pelvic bone. The aitch bone is linked by a ball-and-socket joint to the leg, which runs through the ham.

anatomy—refers to the structure of the animal. It determines how the animal will look, how the animal will move, what tasks the animal will be able to perform, and how sound the animal will be throughout its life.

appearance—the message you convey to a judge through your physical appearance. It is important to remember that although your appearance does not play a role in the competition, it can help you present yourself in a more professional manner during a contest.

art—the skillful use of creative imagination.

B

backfat ratio—is a measure of backfat thickness. Actual backfat thickness is measured with an ultrasound in inches; the measurement is adjusted to 250 pounds live weight.

balance—a term used to describe an animal that has equal portions of width, depth, and length.

binomial system—the formal system for naming species, created by using the genus name followed by the specific name. In writing, the genus name begins with a capital letter while the specific name uses a lower case letter.

blemish—are conditions which cause a noticeable change on an area of the horse but most likely do not impact the way a horse can perform.

blind nipples—nipples that fail to mature and have no opening. These are nonfunctional and create a problem in an animal's ability to mother its young.

breed average—a number that is averaged from production data for animals of a specific breed. This number represents the average or mean number for a certain breed.

breed—a group of domestic animals with a common ancestry and common characteristics that breed true.

breeding program—the planned breeding of a group of animals, managed through the careful selection of desirable traits found in breeding partners.

britch or breech wool—wool from the hindquarters of an animal. This wool is typically the coarsest in crimp and often times looks almost like hair and has a fairly long staple length.

buck knees—a condition where the knee is further forward than it should be in relation to the cannon below it. This can cause structural problems in an animal and is not favorable when judging market or breeding animals.

buckneed—a condition where the animal's knees are bowed or bucked out in front. Since this condition does not allow the legs to be placed squarely and straight beneath the animal, joint wear and pain are more likely.

C

capacity—refers to the width and depth of an animal. Animals with a greater capacity or body volume are preferred to those with a smaller capacity.

carding—the process of brushing raw or washed fibers to prepare them as textiles. Carding is only used for woolen yarn, and in general is done on shorter fibers.

character—refers to the eye catching ability and handling quality of the fleece. This trait can be very difficult to distinguish because it is highly subjective and varies greatly from person to person.

chronological age—refers to the actual age of the animal.

clean weight—is the weight of the fleece after it has been cleaned in a wool testing laboratory. The clean weight is rounded to the nearest 1/10 of a pound.

color—refers to the whiteness and brightness of the fleece. Fleece that is white is more valuable as it represent a greater dying range for the textile industry.

combing—the method for preparing fiber for spinning by use of combs. Combing is only used for worsted yarn, and in general is done on fibers that are long.

comparative terms—vocabulary used in oral reasons to compares two items. These terms allow a presenter to accurately justify the placing and the pairs they found in each class.

concentrate—feeds containing a high percentage of protein and carbohydrates which are generally supplied with grains such as corn for animals.

condition—refers to the amount of fat a breeding animal is carrying. Too little or too much condition can cause problems with fertility and birthing in breeding animals.

conformation—refers to the physical structure and appearance of the animal determined by its skeletal structure.

crimp—a characteristic of wool that tends to lead to the capture of more foreign material in the fleece which reduces the yield.

cuts—a way to determine scores in a judging class. Cuts are assigned to each pair of animals (1 versus 2, 2 versus 3, and 3 versus 4). Small cuts indicate that the two animals being compared are very similar while large cuts indicate more obvious differences between the animals.

D

dew claws—hoofed animals may have a pair of dew claws, or vestigial digits, above their hoofs. In some species they are smaller than the hoofs, and usually never touch the ground, and in others, they are only a little smaller than the hoofs and may touch the ground.

discipline—refers to a particular type or style of riding that a horse may perform.

double muscle—a condition in cattle that arises from breeders selecting animals that display qualities of extreme muscling. These cattle are often undesirable and produce meat that is tough and contains no flavor.

E

efficiency—Optimization of food to growth ratio in animals. Animals with a higher efficiency are able to convert less food to a higher amount of pounds gained.

EPD—Expected progeny difference. This is data that is compiled from an animal's ancestry and is used to predict the differences that can be expected in the offspring of a particular sire or dam over those of other animals used as a reference.

equitation—classes that are judged on the position and effectiveness of the rider. Proper body position of the rider requires that the person sit in the middle of the saddle with weight being evenly distributed on each side of the horse.

expected progeny differences—estimate the genetic value of an animal in passing genetic traits to its offspring. The prediction (EPD) is based on actual performance, progeny performance and relatives' performance.

F

fat free lean gain efficiency PGA—a production characteristic given to an animal that is listed as a number. When listed, this number explains the amount of feed an animal needs to consume for one pound of gain of lean meat. The lower the number, the less feed that is required for one pound of gain of lean meat by the animal.

femur—the thigh bone in an animal. The angle of this bone relative to the ground, allows the ability for animals to freely move.

fleece quantity and quality—four measures of wool that are taken when a sheep is one year of age.

foresaddle—the area of the animal that is comprised of the last rib forward to the base of the neck. The rack contains fairly high-priced cuts, and a grooved shape over the rack would indicate a high degree of muscling.

frame size—refers to the overall height of the animal at maturity; tall animals are larger framed than short animals.

frame—in judging horses this refers to the way the horse carries itself (head carriage, stride length, etc.).

G

gaits—the different ways a horse can move, either naturally or by being trained.

genotype—the genetic makeup of a cell, an organism, or an individual.

grade—refers to the fineness or diameter of the wool fiber and is measured in microns. One micron equals 1/25,400 of an inch.

grease weight—is the weight of the fleece that has been shorn from the sheep when it is one year old. The weight is usually rounded to the nearest 1/10 of a pound and includes all of the material in the wool before it is washed or scoured.

growth size—complete growth size that varies by breed. Frame size is often thought of as just height, when actually it includes body length and body capacity. Frame size can be compared to a rectangular box, with height, length and total volume all making an equal contribution.

growthability—An animal's ability to grow at an acceptable rate. Animals that grow quickly are usually preferred in breeding programs, especially in market animals.

H

halter class—are classes (competitive events) at a horse show where the judge evaluates a horse's conformation based on the job it will likely be performing. There are halter classes for most of the different disciplines of riding.

handling quality—commonly referred to as the finish of an animal. Degree of finish is influenced by the amount of muscling, frame size, and stage of maturity of animal.

hank—is a length of yarn that measures 560 yards long.

head—the head of an animal is the part of the body that houses the brain, eyes, ears, nose and mouth.

hindsaddle—the area of an animal that starts at the last rib and continues to the rear. This contains the most valuable cuts, and is usually a good area to judge muscularity of an animal.

hook bones—the projecting front point of the pelvis on an animal.

hormones—a substance that accounts for the sex characteristics of an animal. The sex characteristics are an indication that the female is producing a large enough quantity of hormones to allow the female to conceive efficiently.

horse judging competition—is a competition in which a person is scored on how accurately they place a class and how well they defend their reasons for placing the class a certain way. A typical horse judging contest has 8 to 12 classes with 4 horses in each class.

horsemanship—the art of riding, handling, and training horses. Good horsemanship requires that a rider control the animal's direction, gait, and speed with maximum effectiveness and minimum efforts.

humerus—the front leg bone of the animal.

hunter over fences—classes that the horse is asked to jump a course of obstacles ranging anywhere from a few inches to 4 foot in height, depending on the class and show specifications. The horse is judged on its jumping style, obedience, and way of going.

hunter under saddle—classes that are judged on the flat (not over jumps) at the walk, trot, and canter. Horses are judged on their way of moving and obedience.

I

index—a type of actual data which allows one animal to be compared with other animals that it was raised with.

inter-muscular fat—fat that is found within the muscle on an animal. This fat produces what is known as marbling, which gives meat flavoring.

inverted nipples—nipples that appear to have a crater in the center and are not functional.

K

keep – cull class—a type of performance data class that consists of eight animals in a group that are both visually evaluated and evaluated based on the performance data. Rather than place them from top to bottom, you are required to pick four from the group to keep as replacements and four to cull from the herd.

keep/cull—a specific class in a livestock judging competition where students must decide which animals to keep and which animals to cull (eliminate from the group) with the assistance of performance data.

kemp—hollow white fibers that will not take dye. This characteristic discounts the fleece and limits its use in the manufacturing process.

KPH—Kidney, Pelvic, and heart fat in an animal. KPH fat is evaluated subjectively and is expressed as a percentage of the carcass weight.

L

lameness—a condition which causes the animal pain or reduces the horse's functionality. Some lamenesses are treatable and the animal is able to return to normal work after a period of time. Some lamenesses, however, are permanent conditions which will always affect the performance of the animal.

lead change—when the horse changes direction during a gait. Types of lead changes include simple lead change and flying lead change.

leg—an appendage of the animal that allows for structural correctness and fluid movement.

ligaments—the structure that connects the bones together.

litter size—the number of pigs that a sow will produce in one litter or pregnancy.

live weight feed efficiency—a production characteristic given to an animal that is listed as a number. When listed, this number explains the amount of feed an animal needs to consume for one pound of gain. The lower the number, the less feed that is required for one pound of gain by the animal.

lock—a piece of wool that is pulled from the fleece for evaluation.

loin eye—the measurement of the muscle that makes up the meat in a pork chop. The size of the loin is used as an indicator for the overall meat in the carcass of the animal.

loin—Area of the animal where muscling is usually measured. A long, wide loin is desirable is breeding and market animals.

longissimus dorsi muscle—the long muscle running next to each side of the backbone. This is the largest muscle in the animal's body and is an indication of the overall amount of muscle in the carcass.

M

marbling—intramuscular fat deposits. These deposits give flavor to the meat of a carcass.

maternal—refers to Indexes and EPDs relating to mothers ability and influence for the offspring. These traits are measured on the sow, regarding farrowing and raising a litter.

maturity—stage of an animal where the growth of bones and muscles ceases, and the animal begins to utilize energy from feed to deposit fat.

MLI—Maternal Line Index. Maternal Line Index is used as a selection tool for producing replacement gilts. MLI combines EPDs for both terminal and maternal traits relative to their economic values in a crossbreeding program, placing twice as much emphasis on reproductive traits as post weaning traits.

monkey mouth—a condition where the lower jaw extends past the upper jaw. This condition is not favorable in animals as it inhibits its ability to consume food properly.

muscling—amount or degree of muscle that an animal carries. This characteristic is important in evaluating animals based on marketing and breeding.

N

natural selection—the process by which favorable heritable traits become more common in successive generations, and unfavorable traits become less common. These traits are usually selected from the strongest or most adaptable animal in the wild.

Number Born Alive (NBA)—is the number of live pigs farrowed in a litter.

number of lambs born—indicates the number of lambs born at one timeEwes may have anywhere from 1 to 3 lambs born at one time. On a rare occasion, they may have as many as 4. Adjusted Weaning weight: a set of adjustment factors that have been calculated to help compare weaning weights from different on a more even basis for lambs.

O

oral reasons—a presentation involving a short oral defense for the way you placed a certain class. This is done from memory and is 2 minutes or less in time.

oral reasons—an impromptu speech given to a judge regarding the placing of a class. This speech will allow the participant an opportunity to explain their reasoning for placing a class, and provides a chance to make up any points lost in the live placings.

P

pale soft and exudative—describing the meat of hogs that are usually bred for heavy muscling. This meat usually contains low amounts of marbling, and when cooked the meat is drying and lacking in taste. Because of these qualities this meat is rejected by the consumer.

parrot mouth—a condition where the upper jaw extends past the lower jaw. This condition is not favorable in animals as it inhibits its ability to consumer food properly.

pasterns—the ankle bones of animal. The angle of the pasterns to the ground is important in animals as it allows them to move freely.

pasterns—the part of an animal that is directly above the hoof. The pastern in vital in absorbing the shock from moving around and can aid animals to move more freely.

performance data—actual data recorded and submitted to breed associations by producers. Producers use performance data to improve carcass, growth and reproductive traits in animals.

performance data—data collected about a particular animal. Two types of production data are actual data, which is taken from the individual animal, and expected progeny difference (EPD), a compilation of data from the animal's ancestry.

phenotype—the physical characteristics or traits of an animal. These traits are often more determined by the environment, than by the genetic makeup of an animal.

physiological age—refers to the way the animal has matured as indicated by bone characteristics, ossification of cartilage, and color and texture of the rib eye muscle.

pigeon footed—a condition found in animals where the front legs point inward. This condition is not favorable in animals as it undermines their structural soundness.

pigeon toed—a condition where the toes or feet turn inward. This condition can affect an animals ability to move freely.

pin bones—rear point of the pelvis on an animal, just below and outside the tail.

pin nipples—are very tiny nipples that are much smaller than the other nipples on the underline, usually they do not function well enough to feed a mother's young.

porcine stress syndrome—a genetic condition passed onto offspring through parents. These pigs often experience increased stress due to some management practices and environments. Pigs with this fatal disease usually experience stress which induces muscle tremors and twitches accompanied with red spots on their underbelly before death.

post legged—a condition where the rear legs are too straight and do not provide enough cushion and flex as the animal walks.

pounds of lean—is a measure of pounds of fat-free lean adjusted to a 185 pound carcass or approximately a 250 pound live pig. The Pounds of Lean is calculated from the backfat & loin eye area EPDs.

presentation—the manner that you present your thoughts and confidence in a competition. Your voice level, mannerisms, and inflection are all aspects of presentation.

production data—data collected for desirable traits in livestock. Traits can include birth weight, weaning weight, or carcass data.

progeny—a general term used to describe the collective offspring.

progeny—refers to the collective offspring of an animal.

purity—a characteristic of wool that is often associated with color but actually it means free from black or brown fibers, kemp, and hair.

Q

quality grade—refers to the eating quality of beef. This is generally determined by two factors: maturity and degree of marbling in the rib eye.

quality—refers to the desirability of an animal. Quality is measured by special characteristics for different animals.

R

rack—a portion of the animal's back that is usually comprised of ribs. In some animals, the rack, is a primal cut of meat.

reasons class—any class in a judging contest where oral reasons will be given following the judging competition. Notes should be taken during this class to help a participant organize their explanation more clearly.

reining—a western performance event that is judged on the horse's obedience and ability to maneuver easily. The horse/rider will perform one of ten patterns in this performance as selected by the competition judge.

reproductive efficiency—an animal's ability to produce offspring in a predictable, regular basis and the young must be born alive and healthy.

retail cuts—muscles of beef that can be cut for the consumer. These cuts of meat are the ones that consumers will purchase, and carcasses are selected for the size of retail cuts that they will produce.

rib eye—the area of the cross section of the *longissimus dorsi* muscle in cattle.

round—the back end of an animal. The round is usually a good way to judge an animal's muscling, and produces many retail cuts from the animal.

rump—the upper rounded part of the hindquarters of an animal. This produces cuts of meat between the loin and the round.

S

scapula—the shoulder blade bone.

scenario—a brief overview of what the producer wants to happen from using the data and the producer's objectives for producing offspring.

scenario—information given to a livestock judging participant in addition to the production data. This should consist of 3 components: the type of production; how the animals are raised; and how the animals are marketed.

science—knowledge acquired through systematic study of understanding how the physical world works.

scouring—the actual cleaning of the wool to remove the dirt, grease, and foreign matter. It is usually done in a lukewarm, mildly alkaline solution, followed by clear water rinses.

sex character—indicates that a bull looks like a male and a heifer looks like a female. Since sex hormones control both the physical appearance of animals and their ability to reproduce, it stands to reason that an animal with more sex character should be more reproductively efficient.

shoulder—a part of an animal where the highest point should connect to the head and neck. The shoulder refers to the scapula and associated muscles.

sickle hocked—a condition where an animal's rear legs are not straight. Because of this condition, undue stress is placed on the stifle muscle, causing the animal to become stifled, that is, the ligament attaching the stifle muscle tears.

soundness—is a term that describes an animal's physical health and condition. It is generally used to describe whether the animal has injuries which prevent it from being able to function properly.

SPI—Sow Productivity Index. Sow Productivity Index is an index that ranks individual animals for reproductive traits.

spinning count—a method of grading wool that refers to the number of "hanks" of.

splay footed—a condition where the front feet of an animal point in opposite directions. This condition is not desirable because excess stress is placed on the knees and joint wear can lead to discomfort to the animal.

splayfooted—a condition characterized by abnormally flat and turned-out feet.

stable length—is the actual length of the wool fiber. The length is taken without the fiber being stretched out and is rounded to the nearest 1/10 of an inch.

STAGES—The Swine Testing And Genetic Evaluation System. This swine data management system was founded in 1985 through Purdue University, the USDA, the National Association of Swine Records, and the National Pork Producers Council.

staple length—a category used to describe the length of wool. This determines primarily which system may be used to spin the fibers into yarn.

strong top—the top line refers to the length of the animals back. An animal's back should be strong and level for more flexibility and movement.

structural soundness—encompasses an animal's skeletal system and how well the bones support the animal's body. These bones make up the skeletal system and are the frame upon which the muscles and internal organs are suspended.

structure—the anatomy, or makeup, of an animal.

SWAPSM—Swine Welfare Assurance Program. An education and assessment tool produced by the National Pork Board, designed to promote and continually enhance swine welfare on the farm.

T

terminal—refers to the intention of the producer for the animals to be harvested and not to be used as breeding stock.

top strength—refers to the strength and muscling of the animals back. An animal's back should be strong and level for more flexibility and movement.

TQA—Transport Quality Assurance. This guideline was established by the National Pork Board, where truckers complete a program to pledge to transport pork with the highest regard for safety and quality of the animal.

twist—the area where an animal's rear legs join. Animals with muscling should have a distinct rivet with a crease where the quarters join.

type—refers to how well an animal represents its particular breed. Most breeds have unique qualities by which they can be identified.

U

uniformity—refers to the variation in grade and length of the fleece.

V

vulva—a reproductive part of an animal that is important to the mating process. The size of a female animal's vulva can indicate the size of the entire reproductive tract.

W

weight—the yield of wool shorn from a sheep at one time. The more foreign matter apparent in the fleece, the higher the weight, but lower the yield.

western riding—a unique event where the horse is judged on its movement, obedience, and ability to perform technical maneuvers. The horse is required to perform a specified pattern that includes flying lead changes, agile turning, and negotiating over a log.

woolen yarn—a manufacturing system within apparel wool that uses some non-processed wool but also contains short fiber by-products from the worsted mills (noils), wool waste, and recycled yarn. This wool is mainly used for blankets, overcoats, and women's suits.

worsted yarn—a manufacturing system within apparel wool that selects for wool that has not been previously processed. This wool is mainly used for men's suits. Yarn that can be spun from one pound of wool top.

Y

yield grade—refers to the amount of lean retail cuts that will come from a carcass. Yield grade is denoted by numbers 1 through 5, with Yield Grade 1 representing the highest cutability or yield, of closely trimmed retail cuts.

yield—the percentage of clean wool fibers present in a fleece that has just been sheared. It does not take into account washing and scouring of the fleece.

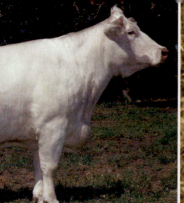

INDEX